Leonid Perlovsky, Ross Deming, and Roman Ilin

Emotional Cognitive Neural Algorithms with Engineering Applications

Studies in Computational Intelligence, Volume 371

Further volumes of this series can be found on our homepage: springer.com

Vol. 349. Vitor R. Carvalho
Modeling Intention in Email, 2011
ISBN 978-3-642-19955-4

Vol. 350. Thanasis Daradoumis, Santi Caballé, Angel A. Juan, and Fatos Xhafa (Eds.)
Technology-Enhanced Systems and Tools for Collaborative Learning Scaffolding, 2011
ISBN 978-3-642-19813-7

Vol. 351. Ngoc Thanh Nguyen, Bogdan Trawiński, and Jason J. Jung (Eds.)
New Challenges for Intelligent Information and Database Systems, 2011
ISBN 978-3-642-19952-3

Vol. 352. Nik Bessis and Fatos Xhafa (Eds.)
Next Generation Data Technologies for Collective Computational Intelligence, 2011
ISBN 978-3-642-20343-5

Vol. 353. Igor Aizenberg
Complex-Valued Neural Networks with Multi-Valued Neurons, 2011
ISBN 978-3-642-20352-7

Vol. 354. Ljupco Kocarev and Shiguo Lian (Eds.)
Chaos-Based Cryptography, 2011
ISBN 978-3-642-20541-5

Vol. 355. Yan Meng and Yaochu Jin (Eds.)
Bio-Inspired Self-Organizing Robotic Systems, 2011
ISBN 978-3-642-20759-4

Vol. 356. Slawomir Koziel and Xin-She Yang (Eds.)
Computational Optimization, Methods and Algorithms, 2011
ISBN 978-3-642-20858-4

Vol. 357. Nadia Nedjah, Leandro Santos Coelho, Viviana Cocco Mariani, and Luiza de Macedo Mourelle (Eds.)
Innovative Computing Methods and their Applications to Engineering Problems, 2011
ISBN 978-3-642-20957-4

Vol. 358. Norbert Jankowski, Włodzisław Duch, and Krzysztof Grąbczewski (Eds.)
Meta-Learning in Computational Intelligence, 2011
ISBN 978-3-642-20979-6

Vol. 359. Xin-She Yang, and Slawomir Koziel (Eds.)
Computational Optimization and Applications in Engineering and Industry, 2011
ISBN 978-3-642-20985-7

Vol. 360. Mikhail Moshkov and Beata Zielosko
Combinatorial Machine Learning, 2011
ISBN 978-3-642-20994-9

Vol. 361. Vincenzo Pallotta, Alessandro Soro, and Eloisa Vargiu (Eds.)
Advances in Distributed Agent-Based Retrieval Tools, 2011
ISBN 978-3-642-21383-0

Vol. 362. Pascal Bouvry, Horacio González-Vélez, and Joanna Kolodziej (Eds.)
Intelligent Decision Systems in Large-Scale Distributed Environments, 2011
ISBN 978-3-642-21270-3

Vol. 363. Kishan G. Mehrotra, Chilukuri Mohan, Jae C. Oh, Pramod K. Varshney, and Moonis Ali (Eds.)
Developing Concepts in Applied Intelligence, 2011
ISBN 978-3-642-21331-1

Vol. 364. Roger Lee (Ed.)
Computer and Information Science, 2011
ISBN 978-3-642-21377-9

Vol. 365. Roger Lee (Ed.)
Computers, Networks, Systems, and Industrial Engineering 2011, 2011
ISBN 978-3-642-21374-8

Vol. 366. Mario Köppen, Gerald Schaefer, and Ajith Abraham (Eds.)
Intelligent Computational Optimization in Engineering, 2011
ISBN 978-3-642-21704-3

Vol. 367. Gabriel Luque and Enrique Alba
Parallel Genetic Algorithms, 2011
ISBN 978-3-642-22083-8

Vol. 368. Roger Lee (Ed.)
Software Engineering, Artificial Intelligence, Networking and Parallel/Distributed Computing 2011, 2011
ISBN 978-3-642-22287-0

Vol. 369. Dominik Ryżko, Piotr Gawrysiak, Henryk Rybinski, and Marzena Kryszkiewicz (Eds.)
Emerging Intelligent Technologies in Industry, 2011
ISBN 978-3-642-22731-8

Vol. 370. Alexander Mehler, Kai-Uwe Kühnberger, Henning Lobin, Harald Lüngen, Angelika Storrer, and Andreas Witt (Eds.)
Modeling, Learning, and Processing of Text Technological Data Structures, 2011
ISBN 978-3-642-22612-0

Vol. 371. Leonid Perlovsky, Ross Deming, and Roman Ilin
Emotional Cognitive Neural Algorithms with Engineering Applications, 2011
ISBN 978-3-642-22829-2

Leonid Perlovsky, Ross Deming, and Roman Ilin

Emotional Cognitive Neural Algorithms with Engineering Applications

Dynamic Logic: From Vague to Crisp

Springer

Authors

Dr. Leonid Perlovsky
Visiting Scholar, Harvard University
Principal Research Physicist and
Technical Advisor
Air Force Research Laboratory 80 Scott Drive,
Hanscom AFB, MA 01731-2909
USA
Tel.: 781-377-1728; Fax: 781-377-8984
E-mail: Leonid.Perlovsky@hanscom.af.mil
leonid@seas.harvard.edu

Dr. Ross Deming
Consultant
Air Force Research Laboratory 80 Scott Drive,
Hanscom AFB, MA 01731-2909
Sensors Directorate
USA
E-mail: Ross.deming@hanscom.af.mil

Dr. Roman Ilin
Computer Scientist,
Air Force Research Laboratory,
Sensors Directorate,
80 Scott Drive Hanscom AFB, MA 01731
USA
E-mail: Roman.ilin@hanscom.af.mil

ISBN 978-3-642-22829-2 e-ISBN 978-3-642-22830-8

DOI 10.1007/978-3-642-22830-8

Studies in Computational Intelligence ISSN 1860-949X

Library of Congress Control Number: 2011934156

Typeset & Cover Design: Scientific Publishing Services Pvt. Ltd., Chennai, India.

Printed on acid-free paper

9 8 7 6 5 4 3 2 1

springer.com

Preface

DL was developed in our research group over the past 15, or so, years. The book disseminates this breakthrough mathematical-engineering idea, which results in 100 times improvement and better in classical algorithmic areas that have been intensively studied for decades. Initial developments in DL were described in "Neural Networks and Intellect," by L. Perlovsky, Oxford University Press, 2001 (which is now in the 3rd printing). The current book describes new breakthrough results developed during the last eight years. First we present the basic technique of DL, explain the fundamental mathematical reason why classical techniques in many areas fail for real-world problems, and how DL overcomes this difficulty. We discuss the algorithmic failure of many techniques to reach information-theoretic performance bounds, relate it to computational complexity, and ultimately to the Gödel theory (it turns out that all past algorithms, neural networks, fuzzy systems, used logic at some step and were subject to Gödelian limitations).

Then we describe a number of applications where significant breakthrough improvements were achieved over popular state-of-the-art techniques (detection, clustering, supervised and unsupervised learning, tracking, sensor fusion, prediction, and particularly financial prediction). We follow with novel engineering areas, where revolutionary results were obtained. The theory is extended toward mathematical modeling of the mind, including higher cognitive functions, beyond anything that has been published in engineering books (no competition): mechanisms of the mind-brain (recent neuroimaging experiments proved that brain is actually using DL computations), applications to learning natural language, to language-understanding search engines for the Internet, to modeling interactions between language and cognition, language and emotions, evolution of languages, evolution of cultures, the role of music in evolution of the mind and cultures.

The mind is the best mechanism for solving complex engineering problems. Therefore, it is just natural that developing engineering algorithms and modeling the mind goes hand in hand. Solving complex engineering problems helps understand working of the mind, and cognitively-inspired algorithms work better than classical engineering methods. This approach to engineering is called computational intelligence.

The book is based on about 200 papers published over the last several years describing DL and its applications. Many of them were important events attracting attention and receiving awards. Every book chapter is written anew, all are unified by a common theme – mathematical technique of dynamic logic and by consistent notations. The book is written for students as well as seasoned professionals, it contains details about applications, algorithms, notations, flowcharts, details that are missing in the papers. DL is easy to use as a textbook or manual. Engineering improvements achieved make it stand out over other texts.

The book contains two parallel tracks. First describes DL applications to many existing problems, which can use existing algorithms without much modification. A second track outlines future research directions appropriate for Master and Ph.D. theses. The first track makes up most of the content of the following chapters. The second track is mostly outlined in Problem sections at the end of each chapter.

A widely held opinion about how scientific knowledge accumulates and get accepted assumes that when a novel scientific paradigm appears, which significantly exceeds in performance the previous ones, it gets accepted by scientific and engineering community. By studying historical changes in scientific paradigms, Thomas Kuhn demonstrated that opinion to be naïve, romantic, and wrong. New scientific and engineering discoveries, no matter how much better and widely applicable than the old ones, are accepted only when a generation changes. Most of professors and engineers continue using and teaching the same techniques that they learned when they were young. Now, twenty years after the first publications on DL, we see the beginning of its wide acceptance by researchers, professors, developers, program managers, and customers in many fields.

The book can be used as the main or supplementary text for the following courses: Electrical Engineering, Signal Processing, Applied Probability and Stochastic Processes, Pattern Recognition, System Parameter Estimation, Applied Physics, Computer Science, Control and Dynamical Systems including nonlinear and adaptive control, Bio-inspired Computation, Neural and Cognitive Systems, Language Learning Systems, Cooperative Man-Machine Systems, Modeling of Cultures, Cognitive Engineering, Computational Intelligence. These many diverse areas we covered in a single book by concentrating on the first principles.

List of Common Abbreviations

CC	Combinatorial complexity
CI	Computational Intelligence
DL	Dynamic logic
KI	Knowledge Instinct
l.h.s.	left hand side of an equation
NMF	Neural modeling fields, a neural architecture implementing DL
NN	Neural networks
r.h.s.	right hand side of an equation

Contents

Chapter 1
Algorithmic Difficulties Since the 1950s

We review mathematical approaches to complex engineering problems since the 1950s including rule systems of artificial intelligence, pattern recognition, neural networks, model systems, fuzzy systems. As soon as computers become available in the 1950s, solving complex engineering problems was closely tied with understanding working of the mind. In the 1950s many scientists and engineers were sure that soon computer intelligence would far exceed that of the human mind. It did not happen. In this chapter we consider difficulties faced by algorithms and neural networks designed for modeling the mind and for solving complex problems; we analyze these difficulties and relate them to the fundamental inconsistency of logic discovered by Gödel in the 1930s.

Note, throughout the book, references are not quoted within the text. Instead, they are discussed at the end of each chapter, and listed at the end of the book.

1.1 Short Summary of Early Approaches: Mathematical Difficulties

The first computational approaches to solving complex problems in the 1950s were inspired by known structures of the brain: many interconnected simple processing elements, neurons. The brain can learn to solve problems. Learning, also called adaptation, is based on adaptive properties of neural connections-synapses. In the late 1940s Donald Hebb discovered that synaptic connections grow in strength, when they are used in the process of learning. Intuitively it seemed simple, connect many simple computational elements-neurons and let them solve problems at hand. Many mathematicians and engineers involved in developing learning algorithms and devices were sure that in this way computers would soon surpass by far human minds in their abilities. They call these devices and algorithms neural networks by analogy with neural networks of the brain. One popular algorithm developed by Frank Rosenblatt was called Perceptron. It's main architectural device was a neuron connected to multiple input signals. The connection strengths were adaptive, so that Perceptron could learn. Perceptron, however, could only learn to solve fairly simple problems. In 1969 Marvin Minsky and Seymour Papert published a book that mathematically proved limits to Perceptron learning.

L. Perlovsky et al.: Emotional Cognitive Neural Algorithms with Eng. Appl., SCI 371, pp. 1–13.
springerlink.com

In parallel, statistical approaches to pattern recognition were developed. These methods used two-step approach. First, patterns were characterized by features. Features are combinations of measurements thought to represent essential aspects of patterns, so that they could be differentiated. Features define so called classification space; each feature is a dimension in this space. Second, a classifier is designed. Two popular approaches are used to design a classifier. First is a plane (or more complex surface) in classification space, which separates classes. If D features are used, this is the dimension of the classification space; correspondingly a classifier is a (D-1)-dimensional surface. Second is a variation of a nearest neighbor approach or a kernel method. In this case neighborhoods in a classification space near known examples from each class are assigned to the class. The neighborhoods are usually defined using kernel functions (often bell-shape curves, Gaussians). These methods turned out to be limited by the dimensionality of classification space - how many features one can to use.

The problem with dimensionality was discovered by Richard Bellman (1962), who called it "the curse of dimensionality." The number of training samples had to grow exponentially (or combinatorially) with the number of dimensions. The reason is in the geometry of high-dimensional spaces: there is "no neighborhood", most of the volume is concentrated on the periphery. Whereas kernel functions are defined so that the probability of belonging to a class rapidly falls with the distance from a given example, in high-dimensional spaces volume growth outweigh the kernel function fall.

The methods discussed above are characterized by learning from examples. It turned out that learning is difficult (mathematically insolvable for complex problems). This conclusion was summarized by Marvin Minsky (1965), who suggested that designing learning artificial systems was premature. Newton, he wrote, *learned* Newton laws, but all other scientists read them in books and acquired then ready-made. Human decision making is based not on learning in every case, but on huge amount existing knowledge. Minsky suggested that the first step in artificial intelligence should be based on the similar principles: storing in computers all the relevant knowledge related to a particular set of problems. The most popular systems, practically used until today stored knowledge in a form of "if… then…" rules; e.g. "if cold then turn on heater." These sometimes are called expert systems. Rule-expert systems work efficiently in well defined situations, when every change can be foreseen and planned for. The difficulty is that in the real world there are always many changes, often unpredictable, rules depend on other rules and grow into combinatorially large trees of rules.

In the 1980s model systems were proposed to combine advantages of learning and existing knowledge. Model systems used models with adaptive parameters to represent events or situations. Models accumulated existing knowledge, while parameters could adapt to unpredictable changes. Model systems work in three steps, first, a particular association between models and signals is selected, second, model parameters are fit to these signals according to some criterion, such as maximum likelihood, third, this procedure is repeated for various associations and the best fit is selected. Adaptive models work well in relatively simple situations, when it is possible to consider all relevant combinations among data and models. However, the number of combinations are combinatorially large; for M models and N signals there is M^N combinations. This number grows fast with M and N. In complex situations model systems encounter combinatorial complexity (CC).

In parallel, a second wave of neural network algorithms has been developed. Grossberg studied brain mechanisms since the 1960s. ART neural network emphasized a fundamental principle of brain organization: interaction of bottom-up and top-down neural signals. Learning was a result of interaction between sensor signals (bottom-up) and mental representations (top-down). Werbos developed Backpropagation algorithm, which overcame limitations of Perceptrons. Yet, dozens of developed neural paradigms faced CC of learning. All learning algorithms had to be trained. During training, every object had to be "shown" to a neural network or learning algorithm in all variations (of size, distance, view angles…), but also training had to include these object variations in combinations with any other object that could be around. CC is unsolvable because even a modest number of 100 elements (objects, pixels, samples, etc.) results in 100^{100} combinations; this number is larger than all elementary particle interactions in the entire history of the Universe. No computer ever would be able to learn that many combinations.

1.2 Combinatorial Complexity and Logic

It turned out that combinatorial complexity (CC) encountered in all considered approaches for decades was related to Gödel theory, which many consider the most fundamental mathematical result of the 20th century. We first discuss why a wide diversity of algorithms and neural networks were all bounded by limitations of logic, even so some of the algorithms and neural networks were specifically designed to overcome Gödelian limitations of logic. Than we discuss relationship between CC, logic, the Gödelian theory, and briefly touch on relationships between logic and the mind mechanisms.

Formal logic is based on the "law of excluded middle," according to which every statement is either true or false and nothing in between. Therefore, algorithms based on formal logic have to evaluate every little variation in data or models as a separate logical statement (hypothesis); a large number of combinations of these variations causes CC.

Rule systems were explicitly based on formal logic and encountered logical limitations. Neural networks it seems were not based on logic. The second wave of neural networks developed beginning since the 1980s overcame limitations of Perceptrons and were proven to be capable of unlimited capabilities in several regards. These neural networks were specifically developed to overcome logical limitations of rule systems. However, neural network training procedures were logical, such as: "this is a chair" – a paramount logic statement. According to this logical procedure every training sample has to be presented one by one, and not only individual objects, but their combinations, leading to CC. And, as already mentioned this is a consequence of formal logic inherent in the procedures.

Multivalued logic and fuzzy logic were proposed to overcome limitations related to the law of excluded middle. Yet the mathematics of multivalued logic is no different in principle from formal logic, "excluded third" is substituted by "excluded n+1." Fuzzy logic made a principled step to overcoming the law of excluded middle. However, within fuzzy logic there is no fundamental procedure for determining a degree of fuzziness. Fuzzy systems encountered difficulty related to selecting the "right" degree of fuzziness. If too much fuzziness is specified, the solution does not achieve a needed accuracy, if too little, it will become similar to formal logic. Complex systems require different degrees of fuzziness in various elements of system operations; searching for the appropriate degrees of fuzziness among combinations of elements again would lead to CC. Is logic still possible after Gödel? Bruno Marchal recently reviewed the contemporary state of this field; it appears that logic after Gödel is much more complicated and much less logical than was assumed by founders of artificial intelligence.

Why is complexity of algorithms related to fundamental inconsistency of logic? At the end of the chapter we give references discussing in details relations between the Gödelian theory of logic, Turing theory of computations, and combinatorial complexity of algorithms using logic. Here we give a brief simplified explanation. Gödel proved his theorems by a procedure closely related to Cantor elimination. He proceeds by listing all "decidable" logical statements; that is all statements that could be decided to be true or false. Then Gödel constructs a statement that is *not listed in this complete* list. He constructs this statement by taking diagonal elements of successive statements and altering them. Clearly this new statement will differ from any previous one, it is therefore "undecidable," because it does not belong to a complete list of decidable statements. How does this procedure relate to combinatorial complexity? Gödel considered infinite statements. Let us instead only consider statements of a finite length. There would be only a finite number of different statements of a finite length. So, it seems, we can in principle decide truthfulness or falsity of each statement. Before answering this, let us ask, how many statements we have to evaluate? The total number of statements will be combinatorially large (in terms of the allowed finite length). Therefore, instead of fundamental inconsistency, the result is combinatorial complexity.

1.3 Logic, Aristotle, Alexander the Great, and the Mind

Let us look now from another angle, why so many talented mathematicians believed in logic. The reason is that logic, as many people believe, is the fundamental mechanism of the mind. Is this true? Mechanisms of the mind we consider in chapter 4. Here we briefly look at relationships between the mind and logic. For a long time people believed that intelligence is equivalent to conceptual understanding and reasoning. A part of this belief was that the mind works according to logic. Although it is obvious that the mind is not logical, over the course of the two millennia since Aristotle, and two hundred years since Newton, many people have identified the power of intelligence with logic. Founders of artificial intelligence in the 1950s and 60s, as we mentioned, believed that by relying on rules of logic they would soon develop computers with intelligence far exceeding the human mind.

The beginning of this story is usually attributed to Aristotle, the inventor of logic. However, Aristotle did not think that the mind works logically; he invented logic as a supreme way of argument, not as a theory of the mind. This is clear from many Aristotelian writings, for example, in "Rhetoric for Alexander" Aristotle lists dozens of topics on which Alexander had to speak publicly. For each topic, Aristotle identified two opposite positions (e.g. make peace or declare war; use torture or don't for extracting the truth, etc.). For each of the opposite positions, Aristotle gives logical arguments, to argue either way. Clearly, for Aristotle, logic is a tool to express previously made decisions, not the mechanism of the mind. Logic can only provide deductions from first principles, but cannot indicate what the first principles should be. Logic, if you wish, is a tool for politicians. (Scientists, I would add, use logic to present their results, but not to arrive at these results.) To explain the mind, Aristotle developed a theory of Forms, which will be discussed later. But during the following centuries the subtleties of Aristotelian thoughts were not always understood. With the advent of science, the idea that intelligence is equivalent to logic was gaining grounds. In the 19th century mathematicians turned their attention to logic. George Boole noted what he thought was not completed in Aristotle's theory. The foundation of logic, since Aristotle, was a law of excluded middle (or excluded third): every statement is either true or false, any middle alternative is excluded. But Aristotle also emphasized that logical statements should not be formulated too precisely (say, a measure of wheat should not be defined with an accuracy of a single grain), that language implies the adequate accuracy, and everyone has his mind to decide what is reasonable.

Boole thought that the contradiction between exactness of the law of excluded middle and vagueness of language was at the core of certain mathematical difficulties, and should be corrected. A new branch of mathematics, *formal logic* was born. Prominent mathematicians contributed to the development of formal logic, including George Boole, Gottlob Frege, Georg Cantor, Bertrand Russell, David Hilbert, and Kurt Gödel. Logicians 'threw away' uncertainty of language and founded formal mathematical logic based on the law of excluded middle. Most of scientists today agree that exactness of mathematics is an inseparable part of science, but formal logicians went beyond this. Hilbert developed an approach named formalism, which rejected the intuition as a part of scientific investigation and thought to define scientific objects formally in terms of axioms or rules. Hilbert was sure that his logical theory also described mechanisms of the mind: "The fundamental idea of my proof theory is none other than to describe the activity of our understanding, to make a protocol of the rules according to which our thinking actually proceeds." In the 1900 he formulated famous Entscheidungsproblem: to define a set of logical rules sufficient to prove all past and future mathematical theorems. This entailed formalization of scientific creativity and the entire human thinking.

Almost as soon as Hilbert formulated his formalization program, the first hole appeared. In 1902 Russell exposed an inconsistency of formal procedures by introducing a set R as follows: *R is a set of all sets which are not members of themselves.* Is R a member of R? If it is not, then it should belong to R according to the definition, but if R is a member of R, this contradicts the definition. Thus, either way we get a contradiction. This became known as the Russell's paradox. Its joking formulation is as follows: A barber shaves everybody who does not shave himself. Does the barber shave himself? Either answer to this question (yes or no) leads to a contradiction. This barber, like Russell's set can be logically defined, but cannot exist. For the next 30 years mathematicians where trying to develop a self-consistent mathematical logic, free from the paradoxes of this type. But, in 1931, Gödel has proved that it is not possible, formal logic was inconsistent, self-contradictory.

Belief in logic has deep psychological roots related to functioning of human mind. As we discuss in details in chapter 4, a major part of any perception and cognition process is not accessible to consciousness directly. We are conscious about the 'final states' of these processes, which are perceived by our minds as 'concepts' approximately obeying formal logic. For this reason prominent mathematicians believed in logic. Even after the Gödelian proof, founders of artificial intelligence still insisted that logic is sufficient to explain workings of the mind. We will turn to this throughout the book; for now, let us just state that logic is not a fundamental mechanism of the mind, but the result of mind's operations (in chapter 4 we discuss that dynamic logic gives a mathematical explanation of how logic appears from illogical states).

To summarize, various manifestations of CC are all related to formal logic and Gödel theory. Rule systems rely on formal logic in a most direct way. Self-learning algorithms and neural networks rely on logic in their training or learning procedures: every training example is treated as a separate logical statement. Fuzzy logic systems rely on logic for setting degrees of fuzziness. CC cannot be resolved within logic. Penrose thought that Gödel's results entail incomputability of the mind processes and testify for a need for new physics. An opposite position in this book is that incomputability of logic does not entail incomputability of the mind. CC of mathematical approaches to the mind is related to the fundamental inconsistency of logic. Logic is not the basic mechanism of the mind. CC of algorithms based on logic is related to Gödel theory: it is a manifestation of the inconsistency of logic in finite systems.

1.4 Problems

In problems 1.4.1-1.4.6 you are asked to build various classifiers using the data set given in the following tables. Use the data points in the first table as the training set and the data points in the second table as the testing set.

Training Data Set

Class, h	1	1	1	1	1	1	2	2	2	2	2	2
x_1	0.50	0.60	0.70	0.80	0.90	0.80	0.40	0.60	0.70	0.80	0.9	0.65
x_2	0.10	0.30	0.40	0.45	0.30	0.20	0.15	0.35	0.48	0.55	0.5	0.6

Testing Data Set

Class, h	1	1	1	1	1	1	2	2	2	2	2	2
x_1	0.5	0.6	0.7	0.8	0.9	0.8	0.45	0.6	0.7	0.8	0.9	0.55
x_2	0.12	0.32	0.36	0.1	0.27	0.2	0.12	0.35	0.48	0.6	0.55	0.5

P1.4.1. Rosenblatt's Perceptron consists of a single neuron with hard limiter transfer function (the same as signum function). The input into the neuron is determined as a linear combination of the input vector component. In the case of two dimensional inputs the output of the neuron to the nth input is determined as follows

$$y(n) = sign(w_1 x_1(n) + w_2 x_2(n) + b) = w^T x + b$$

The input is classified as coming from class 1 if y is 1 and class 2 if y is 0. The unknown weight and bias parameters w_1,w_2, and b are determined using the perceptron learning algorithm as follows.

1. Given the training data set {x(i), d(i), i=1..N}, where d(i) =1 or -1
2. Set the initial values of the weight and bias parameters to zero.
3. Select an arbitrary small value 0 < a < 1 and, for each input patter i, define iteration step

w(n)=w(n)+a[d(i)-y(i)]x(i)

b = b + a[d(i)-y(i)]

4. Continue iteration steps until the classification error stops decreasing.

Implement the algorithm above using your favorite programming language and the data sets given above. Use the class indicator d(i)=-1 for data points with class label 2, and d(i)=1 for class label 1. Did you obtain satisfactory classification? Explain your results. Is the data linearly separable?

P1.4.2. Gaussian classifier makes the assumption that each class can be described by multivariate Gaussian probability density. We can estimate the mean and covariance matrix for each class using the standard statistical formulas. The classification is done by computing the likelihood of the data point being classified and using the likelihood ratio test to make the decision.

Determine the mean and covariance of the two classes h=1 and h=2 given above using the following formulas.

$$M(h) = \frac{\sum_{i=1}^{N} x(i)}{N}$$

$$C(h) = \frac{1}{N-1} \sum_{i=1}^{N} (x(i) - M)(x(i) - M)^T$$

For each data point from the testing test, compute the likelihood of the data point for both classes.

$$l(h) = \frac{1}{2\pi\sqrt{|C(h)|}} \exp(-0.5(x - M(h))^T C^{-1}(h)(x - M(h)))$$

The classification decision is based on the ratio l(1)/l(2). If the ratio is greater than 1 the point is assigned to class 1 otherwise to class 2.

Perform the classification described above for all data points in the testing set. Are all the points classified correctly? Explain your results. It may be useful to plot the 2-std ellipses corresponding to each class.

P1.4.3. Nearest neighbor classifier is one of the "memory based" classifiers. The entire training data set is stored in the computer memory and is used for classification. When the new input patter comes in, the distances between this pattern and all of the training patterns are computed. The class label of the closest training data point is selected for classification.

Based on the description above, implement the nearest neighbor classifier for the data given above. Use Euclidean distance. Explain your results.

P1.4.4. Radial basis function (RBF) networks are powerful classifiers. They are based on the idea that a non-linear transformation of the data into a high dimensional space (called feature space) can turn a non-linearly separable problem into a linearly separable one. Thus, the RBF network consists of two layers. The first layer performs the non-linear mapping into the feature space, and the second layer performs the classification in the feature space using linear network similar to the Perceptron. The functions performing the non-linear mapping depend only of the distance between the input pattern and some constant vector called "center", hence the name "radial-basis". The simplest architecture is to use the training data points as centers. This means that the number of neurons will be the same as the number of training data points – a very big number in a realistic scenario. There are ways to address this issue. In this problem, however, we will use this simple architecture since our data set is very small.

Use the following radial basis function for this exercise. Here c is the center and σ is the function parameter influencing the "width" of the function.

$$\varphi(x,c) = \exp(-\frac{(x-c)^2}{\sigma^2})$$

Given N training data points, with corresponding target values d, compute the following N by N matrix

$$\Phi = \begin{pmatrix} \varphi(x(1),x(1)) & \varphi(x(1),x(2)) & \varphi(x(1),x(3)) & \ldots & \varphi(x(1),x(N) \\ \varphi(x(2),x(1)) & \ldots & \ldots & \ldots & \ldots \\ \ldots & \ldots & \ldots & \ldots & \ldots \\ \ldots & \ldots & \ldots & \ldots & \ldots \\ \varphi(x(N),x(1)) & \ldots & \ldots & \ldots & \varphi(x(N),x(N)) \end{pmatrix}$$

The output of the RBF network is determined by the following formula

$$y(x) = \sum_{i=1}^{N} w_i \varphi(x, x(i))$$

The weights can be determined using the condition $\Phi w = d$. Thus

$$w = \Phi^{-1} d$$

Before applying this formula, augment the matrix Φ with an additional column on the right consisting of all ones. The inverse matrix calculation changes to the pseudo-inverse. The last element of the resulting weight vector contains the bias. Remember to use the targets d(i)=-1 for data points with class label 2, and d(i)=1 for class label 1.

Select several values of σ, in the interval [0.1 .. 1]. Classify all the test patterns using the decision rule y(x)>0 -> h=1, otherwise h=2. Are the results of classification satisfactory? Why are they better than the results in 1.4.1-3? Which σ resulted in the best classification?

Write the computer code to draw the decision boundary for this classifier. This is done by computing the value of the classifier y(x) over a 2-dimensional grid letting x_1 and x_2 change from 0 to 1 with a certain fixed step, for example 0.1. The values of the classifier can be displayed using a 2-dimensional scatter plot with color-coded markers (scatter command in MatLab). Draw the decision boundary for various values of σ. How does it influence the decision boundary?

P1.4.5. Support Vector Machine (SVM) is another powerful classifier. It is also based on the idea of non-linear transformation of the input data into the feature space, where the data becomes linearly separable. The basic architecture of the SVM is the same as that of RBF. However, instead of approximating the entire training data set, the SVM focuses on the boundary between the classes trying to find the boundary which best separates the classes by maximizing the "margin" between the boundary and the closest data points. The determination of this boundary involves solving a constrained optimization problem with respect to the output layer weights. To be more precise, it is a quadratic maximization problem with linear constraints. This is equivalent to solving a dual minimization problem with respect to Lagrange multipliers. As we will see, the non-zero Lagrange multipliers correspond to the data points that are the closest to the decision boundary and are called "support vectors". The design of SVM is rather involved and we outline the steps below. We are working with the data sets given in the beginning of Problems section.

Given the data set with N points x(i) and target values d(i). Remember to use the values of 1 and -1 for the targets. Let the non-linear transformation be given by a function $\varphi(x(i))$, as in the case of RBF. We will define another useful function, called the inner-product kernel $K(x_1,x_2)$. This function is defined as follows.

$$K(x_1,x_2)=\varphi^T(x_1)\varphi(x_2)=\sum_{j=1}^{m}\varphi_j(x_1)\varphi_j(x_2)$$

Compute the N by N matrix Q, with elements given by the following formula.

$$Q(i,j)=d(i)d(j)K(x(i),x(j))$$

The quadratic minimization problem is expressed as follows.

$$\begin{cases}\min_a\left[-a^T I+\frac{1}{2}a^T Qa\right]\\ s.t.\\ a^T d=0\\ 0<=a<=C\end{cases}$$

In this problem a is the N by 1 vector of Lagrange multipliers, and C is a constant chosen a priory.

After the problem is solved, the indexes of the non-zero elements of a determine the support vectors. The value of bias b is determined using the following formula, with S denoting the set of support vectors.

$$b = \frac{1}{N_S} \sum_{i \in S} \left[d(i) - \sum_{j \in S} a_j d(j) K(x_i, x_j)) \right]$$

The output of the classifier can be computed in terms of the parameters a and the kernel function as follows.

$$y(x) = \sum_{i=1}^{N} a_i d_i K(x, x(i)) + b$$

The classification of the input x is performed using the usual rule y(x)>0 -> h=1, otherwise h=2. From here we can see the value of introducing the kernel function. We no longer need to use the original transformation $\varphi(x(i))$.

Use the following kernel function

$$K(x_1, x_2) = (1 + x_1^T x_2)^2.$$

Perform all the steps described above: compute matrix Q, solve the minimization problem, and determine the bias. Use your favorite method of solving quadratic optimization problems (for instance MatLab's *quadprog* function, if available). Use C=10. Perform classification of the test data. Describe the performance of this classifier. How does it compare to the other classifiers in problems 1.4.1-4?

Draw the decision boundary of this classifier. This is done by computing the value of the classifier y(x) over a 2-dimensional grid letting x_1 and x_2 change from 0 to 1 with a certain fixed step, for example 0.1. The values of the classifier can be displayed using a 2-dimensional scatter plot with color-coded markers (scatter command in MatLab). How does this decision boundary compare to the decision boundary in problem 1.4.4?

1.4.6. Clustering refers to identifying groups of data points. In this problem assume that the data set has no class labels and apply K-means clustering algorithm with k=2 in order to identify the two classes. K-mean algorithm operates as follows.

1. Select the number of clusters to be discovered, k. In our case k=2
2. Randomly select cluster centers for each cluster, $\mu_k, k = 1,2$
3. Define binary indicator variable $r_{ik}, i = 1..N, k = 1,2$. The value r_{ik} equals to one when data point i is assigned to cluster k.
4. Assign each data point to the closest cluster k, using the following formula

$$r_{ik} = \begin{cases} 1 & ,k = \arg\min_j \left\| x_i - \mu_j \right\|^2 \\ 0 & ,otherwise \end{cases}$$

5. Re-compute the cluster centers using the following formula

$$\mu_k = \frac{\sum_{i=1}^{N} r_{ik} x_i}{\sum_{i=1}^{N} r_{ik}}, \text{ for k=1,2}$$

6. Repeat steps 4 and 5 until the cluster assignments stop changing

Apply the algorithm described above to the data given in the beginning of Problems section. Did the algorithm recover the correct class labels? Explain your results.

Run the algorithm starting from different initial values for the cluster centers. Does the algorithm always converge to the same cluster assignment? Explain your results.

1.5 Literature for Further Reading

1.5.1 Section 1.1, Short summary of early approaches. Mathematical difficulties

Early modeling of neural structures of the brain: McCulloch and Pitts, 1943; McCulloch, 1961; 1965; Hebb, 1949/2002.

Early artificial neural networks: Rosenblatt, 1962.

Mathematical analysis of Perceptron limitations: Minsky & Papert, 1969/1987.

Pattern recognition: Nilsson, 1965; Fukunaga, 1972; Duda and Hart, 1973; Watanabe, 1985.

Neural Networks, second wave: Carpenter & Grossberg, 1987; Grossberg and D. S. Levine, 1987; Grossberg, 1988; Barto, Sutton, & Brouwer, 1981.

Rule systems and AI: Minsky, 1965; 1968; 1975; Winston, 1984; Bonnisone et al, 1991; Keshavan et al, 1993.

Model systems: Nevatia & Binford, 1977; Brooks, 1983; Winston, 1984; Grimson & Lozano-Perez, 1984; Chen & Dyer, 1986; Negahdaripour & Jain, 1991; Bonnisone et al, 1991.

Linguistics: Chomsky' linguistics (inborn Universal grammar); as rule systems: Chomsky, 1959; Koster & May, 1981; Chomsky, 1972; as model systems (principles and parameters): Chomsky, 1981; minimizing number of rules: Chomsky, 1995; an excellent and lively review of the field up to 1991: Botha, 1991.

Combinatorial complexity: Bellman, 1963; Winston, 1984; Segre, 1992; Perlovsky, 1994; 1998.

1.5.2 Section 1.2, Combinatorial complexity and logic

Logic and the mind, logical difficulties: Hilbert, 1928/1967; Russel, & Whitehead, 1908/1962; Gödel, 1986; Marchal, 2005; when analyzing mind: Penrose, 1989; 1994.

Logic and combinatorial complexity (CC): Perlovsky, 1996; 2001; 2006a.

Mathematical analysis of computational models of intelligence: Perlovsky, 1994; 1996; 2001; 2007; 2010c; Also, see Winston, 1984; Simpson, 1990; Girosi, Jones, & Poggio, 1995.

Fuzzy logic: Zadeh, 1965; 1971; Kecman, 2001.

1.5.3 Section 1.3, Logic, Aristotle, Alexander the Great, and the mind.

Models of the mind: Aristotle, 1995; Perlovsky, 2007g; 2010c.

Chapter 2
Dynamic Logic

The strength of logic is in structuring problems according to existing knowledge. Its weakness is absence of dynamics and learning. An opposing approach of "connectivism" or neural networks is dynamical and has been conceived to be capable of learning. Its weakness is that it cannot easily incorporate structural knowledge. As discussed in the previous chapter, both approaches faced combinatorial complexity (CC). DL combines structure and dynamics, ability to utilize prior knowledge and ability to learn. In this way it is similar to model-based approaches. DL fits models to data, while avoiding combinatorial complexity (CC) of the past algorithms and neural networks. DL can be considered as a gradient ascent along variables, which used to be considered as essentially discrete; DL makes discrete variables into continuous and also avoids local maxima. Another way of viewing DL is as a modification of fuzzy logic such that degrees of fuzziness for various models are autonomously updated and reduced along with improved accuracies of the models. DL is a process; its initial state is a vague-fuzzy state (model) in which vagueness corresponds to the uncertainty of knowledge (inaccuracies of models). This DL process "from vague-to-crisp" corresponds to the Aristotelian conception of forms evolving from illogical forms-as-potentialities to logical forms-as actualities. In the DL process vagueness decreases, while models become more similar to patterns in data. The number and types of models are also adjusted to improve the similarity between models and data. In this chapter we define similarity measures, DL process equations, and discuss DL convergence.

2.1 Similarity Measure between Models and Data

Learning algorithms often maximize a similarity between incoming signals and an algorithm's internal representation of the world. A similarity, in this way, is an algorithm's measure of knowledge of the world. And principal differences among algorithms are in how they represent this knowledge. Model-based (or just model, for shortness) learning algorithms maximize a similarity between incoming signals and internal models (of the world processes and events). DL is a version of a model learning. Reinforcement learning maximizes reward value by mapping states of the world into actions; some reinforcement learning algorithms can be formulated as maximizing knowledge. The measure of knowledge (or reward value) might be internal to an algorithm, such as in model learning and in reinforcement learning, or external, such as in supervised learning.

L. Perlovsky et al.: Emotional Cognitive Neural Algorithms with Eng. Appl., SCI 371, pp. 15–30.
springerlink.com

Supervised learning uses training data, which consists of explicit pairs of examples of input signals and desired output (output numbers or categories, or classes). Supervised learning can be used to estimate a function that maps input signals to desired outputs (this is similar to, maximization of knowledge). The knowledge function can consist of rules, local in the space of signals, such as the nearest neighbor algorithms (in this popular class of algorithms future decisions are made similar to the past, training experience). Another popular technique of supervised learning is support vector machines (SVM) and the underlying Statistical Learning Theory; this technique emphasizes that no internal representation of knowledge is necessary (this is true about initial "canonical" formulation; recent formulations explore advantages of internal representations of knowledge).

DL is an unsupervised model-based learning technique (DL and model learning, in general, can be used in a supervised setting; this makes learning problem much simpler and limits. Three questions are essential for model-based algorithms. First, where are models coming from? In several following chapters we discuss how general models are designed for several classes of problems. The more general approaches to designing algorithms, which learn to construct models on their own, we discuss near the end of the book, when we look at what is known at how the human mind does it. Second, how to construct a similarity measure? Simple similarity measures, such as least mean square are appropriate for simple problems, e.g. linear regression; but they cannot solve complex problems, e.g., when several classes of data should be learned. Complex similarity measures, appropriate for complex multi-class problems, often lead to combinatorial computational complexity, which we have discussed in chapter 1. Below we consider a fairly general similarity measure of this type. The third question, essential for model-based techniques, is how to maximize such a general similarity measure, while avoiding combinatorial complexity. This problem has been mathematically difficult, because it includes *association* of models and input signals. Solution to this problem is an essence of DL, and we consider its mathematical formulation in this chapter.

Models in DL we denote $\mathbf{M}_m(\mathbf{S}_m,n)$; we enumerate models by index m = 1,… M (please note, we use bold **M** for the models and regular M for the total number of models; at first it might seem cumbersome, but years of experience proved these notations useful). Each model is characterized by its parameters, $\mathbf{S}_m$; which are generally unknown, and learning consists in estimating these model parameters, as well as the number of models M. Each model predicts expected values of a signal in a sample number n; a single model usually predicts signals in many samples. Associating models with corresponding signals (m with n), as mentioned, has been traditionally the most difficult mathematical part of model learning; in other words, a part of learning includes deciding which signal n "comes" from which model m. Signals, $\mathbf{X}(n)$, are enumerated by index n = 1,… N. Index n might be characterized by geometric and time coordinates (which are not included in the list of model parameters, if known). Signals $\mathbf{X}(n)$ and models $\mathbf{M}_m$ are often multidimensional quantities, vectors, and we denote vectors by bold.

As following chapters demonstrate, a powerful similarity measure between a set of input signals {**X**} and models {**M**}, suitable for many applications, can be defined as follows

$$L(\{\mathbf{X}\},\{\mathbf{M}\}) = \prod_{n\in N} \ell(\varepsilon(n)). \tag{2.1.1}$$

Here Π denotes a product over index n=1,... N, ε(n) is an error between a signal **X**(n) and models {**M**}, defined under a multi-modal assumption; namely, that this signal could come from any of model m=1,... M, with certain similarity appropriate :

$$\ell(\varepsilon(n)) = \sum_{m\in M} r_m \, \ell(\varepsilon(n)|m), \tag{2.1.2}$$

Here, $\ell(\varepsilon(n)|m)$, or $\ell(n|m)$ for shortness are conditional similarities; notations (n|m) are read "n given m." They are defined similar to conditional probability density functions (pdf), or likelihoods. If the models, parameter values, and pdf functional forms are correct, then $\ell(n|m)$ are indeed conditional likelihoods, L is a total likelihood, and maximization of (2.1-1) over the parameters yields the maximum likelihood estimation of the parameters. In correspondence with this probabilistic analogy, conditional similarities are defined under an assumption that one object m is present (and normalized like standard pdf, $\int \ell(\varepsilon(n)|m)d\varepsilon(n)=1$). The actual number of objects m being present is characterized by a parameter r_m,

$$r_m = N_m/N, \tag{2.1.3}$$

representing the ratio of the number of objects type m, N_m, to the total number of objects, N. Corresponding coefficients in statistics are called priors; in this book we usually call them rates. In general, r_m are not known and have to be estimated along with other parameters. According to the definition,

$$\sum_{m\in M} r_m = 1. \tag{2.1.4}$$

Combining (2.1-1) and (2.1-2) we obtain

$$L = \prod_{n\in N} \left[\sum_{m\in M} r_m \, \ell(n|m)\right]. \tag{2.1.5}$$

This expression, if one opens brackets and multiplies the items, contains total of M^N items. Each item corresponds to a particular association between data {**X**(n)} and models {$\mathbf{M}_m$}; expression (2.1-5) contains all possible associations. The larger than astronomical, combinatorial, number, M^N, explains CC of many algorithms. For example, multiple hypothesis testing, which is at the core of many algorithms, attempts to maximize similarity L over model parameters and associations between signals and models, in two steps. First it takes one of the M^N items, which is one particular association between signals and models; and maximizes it over model parameters. Second, the largest item is selected (that is the best association for the best set of parameters). Such a program inevitably faces a wall of CC, the number of computations on the order of M^N. This is a logical way of maximizing (2.1-5), and it explains why logic cannot solve this problem. DL maximizes (2.1-5) without combinatorial complexity, as we discuss in the next section.

2.2 DL Process from Vague to Crisp

Let us briefly recollect discussion from chapter 1 as related to the previous section. Logic leads to CC because it attempts to maximize similarity (2.1-5) item by item. Fuzzy logic could speed up calculations at the expense of exactness, if one knows what is an appropriate fuzziness for each item. But different items in (2.1-5) might require different degrees of fuzziness, and sorting through degrees of fuzziness again would lead to CC. A simple attempt to use gradient ascent and to modify parameters according to the gradient of similarity (2.1-5) would not work, because (2.1-5) is a highly nonlinear function; a maximum of every item in (2.1-5) is a local maximum, therefore trying to cope with local maxima would again lead to CC. DL combines the idea of gradient ascent with Aristotelian suggestion that this process has to move from an illogical solution to a logical one, in other words it should start with a fuzzy solution. Mathematically, "starting fuzzy," smoothes out local maxima, and opens a door to fast gradient-like solution.

An important aspect of DL is matching vagueness or fuzziness of similarity measures to the uncertainty of models. Initially, parameter values are not known, and uncertainty of models is high; so is the fuzziness of the similarity measures. In the process of learning, models become more accurate, and the similarity measure more crisp; the value of the similarity increases. This is the mechanism of dynamic logic. Mathematically it is described as follows. First, assign any values to unknown parameters, $\{\mathbf{S}_m\}$. Then, compute association variables f(m|n),

$$f(m|n) = r_m \, \ell(n|m) \,/ \sum_{m' \in M} r_{m'} \, \ell(n|m'). \tag{2.2.1}$$

Whereas $\ell(n|m)$ may vary between 0 and infinity, f(m|n) vary between 0 and 1. Therefore f(m|n) are more convenient for understanding DL. Equation (2.2-1) looks like the Bayes formula for a posteriori probabilities; if $\ell(n|m)$ are conditional likelihoods, f(m|n) are Bayesian a posteriori probabilities for signal n originating from object m. According to the probabilistic analogy, the following is a standard terminology: $\ell(n|m)$ is a measure (likelihood) that data **X**(n) are observed given that they came from an object described by model m; and f(m|n) is a measure that the object m is a source of data **X**(n), given that these data were observed (on the first reading this standard terminology might sound a bit tautological; do not try to "get" it on the first reading, it is really a simple staff and will come naturally a chapter later).

The DL is a process of maximizing similarity (2.1-5) over the parameters $\{\mathbf{S}_m, r_m\}$ and the number of models, M. Conditional likelihoods may have their own parameters in addition to parameters of model, we consider them included into vectors $\mathbf{S}_m$. In problems to this chapter we consider derivation of the DL equations, which we state below. DL is defined by the following differential equations (variable t, with respect to which the derivatives are taken in the left hand side (l.h.s.) of equations below, is an internal time of convergence of the DL process, which is determined by a computer speed, not by external processes in real time),

$$df(m|n)/dt = f(m|n) \sum_{m' \in M} [\delta_{mm'} - f(m'|n)]\, [\partial \ln \ell(n|m')/\partial \mathbf{M}_{m'}]\; \partial \mathbf{M}_{m'}/\partial \mathbf{S}_{m'}\; d\mathbf{S}_{m'}/dt, \tag{2.2.2}$$

$$d\mathbf{S}_m/dt = \sum_{n \in N} f(m|n)[\partial \ln \ell(n|m)/\partial \mathbf{M}_m]\partial \mathbf{M}_m/\partial \mathbf{S}_m\,; \tag{2.2.3}$$

$$dr_m/dt = \sum_{n \in N} f(m|n)[1/r_m] + \lambda(\sum_{m' \in M} [r_{m'} - 1]\,); \tag{2.2.4}$$

Here, $\delta_{mm'}$ is 1 if m=m', 0 otherwise; $\ln \ell(n|m')$ is a natural logarithm of the corresponding conditional similarity $\ell(n|m')$. The last derivative in the r.h.s in (2.2-2) is given by (2.2-3), which is substituted into (2.2-2). In eq. (2.2-4) λ is a so-called Lagrange coefficient, which is found so that r_m satisfy condition (2.1-4). Maximization over the number of models M we consider later.

Eq.(2.2-3) is similar to the gradient ascent, parameters $\mathbf{S}_m$ are modified so that similarity (2.1-5) increases on each time step. A difference from the regular gradient ascent is in coefficients-weights f(m|n). They associate gradients from every point n to every model m. The problem of association of data and models used to be considered essentially discrete: previous model-based algorithms considered associations of every data point n and model m as 1 or 0 (associated or not-associated). DL turned these discrete variables into continuous f(m|n). And "gradient assent" along parameters, eq.(2.2-3), is accompanied by a "gradient ascent" along associations f(m|n), eq.(2.2-2). Actually these two together make real gradient ascent, similarity L increases at every time step.

This paragraph may be omitted; it is only needed for those who seek a deeper understanding. Considering associations as continuous variables, similar to probabilities, seems so natural that it is not clear why it was not widely accepted decades ago (given the fact that it results in solutions of previously unsolvable problems), and what a big deal is about DL. The next chapter 3 considers many applications of this idea, and every one seems completely natural and the obvious way to go. The problem section in chapter 3 takes willing readers through previously used algorithms (still currently popular); then it might become clearer why this step has been counterintuitive and has taken decades in algorithm development, and why many people still solve problems using discrete associations. Actually, using continuous associations between data points and object models is just a first step. Consider continuous association between words and sentences. Try to define a continuous measure of associating, say, this word "say" to this current sentence; define it in such a way that one can take a gradient along this association to decide if this word belongs to this sentence. That next step in *making associations continuous* may reveal its counterintuitive complexity. It is essential for learning language (phrases as composed of words, paragraphs of phrases), and for learning situations around us as composed of objects. We consider these problems near the end of chapter 3 and in chapter 4, and it will become obvious and natural how to define continuous associations among words and sentences. In chapter 4 we also discuss evidence that DL actually models mechanisms of the human brain. Let us now return from heavy thinking to simple math. equations.

Equations (2.2-2, 2.2-3, 2.2-4) are linear differential equations of the first order with condition (2.1-4) and they can be solved by any differential equation solver in a straightforward way. This of cause requires that conditional similarities $\ell(n|m)$ and models $\mathbf{M}_m$ are specified as functions of their parameters. Also the initial values of parameters must be chosen.

The following chapters will consider a number of specific similarities and models for various applications. In simple cases models $\mathbf{M}_m$ are given by straightforward equations, in complex cases, models $\mathbf{M}_m$ could be given numerically, e.g. as solutions of electromagnetic, hydrodynamic, or financial equations. One should keep in mind that in either case the r.h.s. of these equations can be computed numerically, including taking the derivatives. A simplicity or complexity of models in the r.h.s. is a separate problem from using the DL process, given by the above equations for maximization of similarity (2.1-5). One can write once a computer code for solving the above equations, in which similarities and models are input quantities, and than apply it to any problem, where similarity (2.1-5) should be maximized. In this way, the DL process given by the above equations is a simple and general principle. It can be applied to solving many problems, which have been unsolvable due to CC, and several of these problems we consider in the following chapters.

Instead of using a standard differential equation solver, one can write his/her own code to solve these equations. For this purpose, first, multiply these equations by dt. Second, consider the l.h.s. of the equations as a change, df and $d\mathbf{S}_m$, in the corresponding quantities, f and $\mathbf{S}_m$ from iteration number (it) to iteration (it+1), df = $f^{it+1}(m|n)$ - $f^{it}(m|n)$ and $d\mathbf{S}_m = \mathbf{S}_m^{it+1} - \mathbf{S}_m^{it}$. Eq.(2.2-1) can be used instead of (2.2-2), with r.h.s. computed using parameters from the previous iteration. Now we can rewrite these equations as iterative equations:

$$f^{it+1}(m|n) = [\, r_m \ell(n|m) / \sum_{m' \in M} r_{m'} \ell(n|m')\,]^{it}, \qquad (2.2.5)$$

$$r_m^{it+1} = (1/N) \sum_{n \in N} f(m|n). \qquad (2.2.6)$$

$$\mathbf{S}_m^{it+1} = \mathbf{S}_m^{it} + dt \cdot \sum_{n \in N} f(m|n)[\partial \ln\ell(n|m)/\partial \mathbf{M}_m]\partial \mathbf{M}_m/\partial \mathbf{S}_m; \qquad (2.2.7)$$

Eq (2.2-6) is equivalent to (2.2-4) and satisfies condition (2.1-4). To solve these equations, one follows the algorithm:

(1) start with iteration number it = 1,
(2) select initial values of the step size dt and parameters according to existing information (if there is no information, this selection could be random, dt should be small enough as considered later),
(3) compute the r.h.s., and
(4) obtain improved values of f, $\mathbf{S}_m$, and r_m for the next iteration. On the iteration it = 2, these improved values of parameters are used to compute the r.h.s., and iterations continue until convergence (until parameters stop changing significantly on the next iteration).

In some cases of simple models the derivatives in the r.h.s. can be taken analytically, simplifying the equations. This fact is secondary; using a general computer code and numerical procedures for computing the r.h.s. is a general procedure that can be used throughout this book. However, working with simplified equations, may add an intuitive light for the DL process; this is useful for understanding of how exactly to apply DL to new applications, how to select conditional similarities, models, the step size dt, and initial model parameters to start the solution process, and how to diagnose errors in computer codes. Therefore, we use both types of procedures in the book, and share the know-how, gradually accumulated since the late 1980s, how to select these quantities, how to diagnose computer errors and other problems.

To summarize what is demonstrated in the following chapters in many examples, a typical DL process defined by eqs. (2.2-2, 2.2-3, 2.2-4) or by (2.2-1, 2.2-6, 2.2-7) begins with large uncertainties, corresponding to incorrect values of parameters. Incorrect models do not match the data, all standard deviations (std) are large, all data points have small but non-zero similarities with all models, all f(m|n) are flat, small but non-zero. In the course of iterations, parameter values improve, models better fit patterns in data, std tend to small values corresponding to sensor errors, f(m|n) tend to zeroes or ones, so that data points $\mathbf{X}(n)$ are assigned to their models. In case of uncertainties in data, due to large sensor errors, or due to natural overlap in data patters, f(m|n) deviate from 0 or 1, corresponding to data properties.

2.3 Mutual Information Similarity for Approximate Models

Similarity (2.1-5) is analogous to probabilistic likelihood measure. Its theoretical ground is firm, when for some values of parameters conditional similarities are likelihoods (in this case, similarity tells how "likely" is to observe given data; also using likelihoods leads to certain theoretical optimalities). However, when models are approximate at best, this interpretation has no grounds. Instead, here we consider a modification that can be interpreted on the information ground. When models are approximate, it makes sense to extract maximum information from data, which could be extracted using available knowledge-models. A similarity based on mutual information between models and data is given by

$$LL = \sum_{n\in N} abs(\mathbf{X}(n)) \bullet \ln [\sum_{m\in M} r_m \ell(n|m)]. \quad (2.3.1)$$

The only difference from similarity (2.1-5) is a weight abs(**X**(n)); every sample n is weighted with the strength of signal in this sample. This similarity measure attempts to match partial similarities to the signal strength, abs(**X**(n)). In the literature section at the end of this chapter we reference detailed discussions, why this modification can be interpreted under certain conditions that LL is mutual information between a set of data {**X**}, and a set of models {**M**}, even if models are approximate. *Therefore maximization of similarity is interpreted as maximization of information in the model about the data.* **The only changes to the DL equations from the previous section are that all sums over n, in eqs.(2.2-3)** through **(2.2-7), are substituted with weighted sums** using abs(**X**(n)) weights, so instead of 2.2-3, 2.2-4, 2.2-5, we should use

$$f^{it+1}(m|n) = [r_m \ell(n|m) / \sum_{m'\in M} r_{m'} \ell(n|m')]^{it}, \quad (2.3.2)$$

$$r_m^{it+1} = (1/N) \sum_{n\in N} abs(\mathbf{X}(n))f(m|n). \quad (2.3.3)$$

$$\mathbf{S}_m^{it+1} = \mathbf{S}_m^{it} + dt \cdot \sum_{n\in N} abs(\mathbf{X}(n))f(m|n)[\partial \ln\ell(n|m)/\partial \mathbf{M}_m]\partial \mathbf{M}_m/\partial \mathbf{S}_m . \quad (2.3.4)$$

And instead of (2.1-4), r_m satisfy the following condition

$$\sum_{m\in M} r_m = (1/N) \sum_{n\in N} abs(\mathbf{X}(n)). \quad (2.3.5)$$

Note that in section 2 signals were separate individual data points, as happens when a continuously measured signal is compared to a threshold, and signals below the threshold are discarded. Here signal X(n) is present at every value of its index n, which is a usual case for images.

2.4 Number of Models

A typical procedure for selection the number of models, M, involve solving the DL equations for several model numbers and selecting the one that maximizes similarity L. If M is approximately known, one can try all expected values of M. A more efficient procedure, when M is not known, solves the DL equations for 1 model, for 2 models, etc, until similarity L continues increasing. Even *more efficient* version of this procedure is as follows. During DL iterations (considered in the previous section) always keep a "dormant" model, in addition to actively adapted-estimated models. For a dormant model std is kept large, on the order of $(X_{max}-X_{min})$ for each dimension of **X**, and parameters are not updated from iteration to iteration, except for r_m. If r_m exceeds a predetermined threshold, this indicates that a new meaningful pattern of data is tentatively detected. Then the dormant model is activated (all its parameters are adapted on each iteration), and a new dormant model is initiated. If there are no considerations suggesting a threshold value, the threshold can be included in the list of adaptive parameters. If several functionally different types of models are used, a dormant model should be kept for each type.

A larger number of models could always fit data better; even if improvement is superficial, still, the value of L can be increased if more models are used (for example, an extra model can be used to describe any one data point very accurately, leading to increase of L). Therefore often a penalty function should be introduced to correct for this. Similarity L is multiplied by a penalty function to reduce it for expected "superficial" improvement. To establish what is superficial is not trivial. Let us consider several types of penalty functions.

A well known penalty function is called *Akaike Information Criterion (AIC)*. It is based on the following considerations. Consider logarithm of similarity,

$$LL = \ln L. \tag{2.4.1}$$

The average LL value is proportional to a number of data points. Consider average similarity per data point, LL/N; and consider a case, when a correct number of models exists, so average value of LL/N has its true value. Ideally maximizing similarity should lead to this true value on average. However, since more models can always be used to increase LL/N beyond its true value, the penalty function should be used to compensate for this superficial increase. Such a penalty function was estimated for the case, when L is likelihood, and the number of data points $N \to \infty$. In this case, an amazingly general result has been obtained by Akaike.

Let us make a short step aside to use statistically correct terminology (a reader who does not care about the difference between an average and expected value can skip this paragraph, it is not essential for much of the content of this book). Statisticians often consider so called "expected value" of a statistical variable (for x it is denoted $E\{x\}$), which can be considered an idealized notion of an average. Whereas an average value is computed by averaging data, an expected value is a theoretical notion, which is obtained by "averaging" the variable with its pdf, not with data. This procedure is called "taking an expected value." It is assumed that an average value -> to expected value, when $N \to \infty$.

Akaike proved that the expected value of LL/N, when it is estimated from data using varying number of models, depends only on the total number of parameters in all models. It does not depend on functional shapes of models or conditional likelihoods! The difference, called bias in statistics, between the true value of LL/N and the expected value of LL/N estimated from data using p parameters, when $N \to \infty$, is

$$\text{Akaike bias, defined as } [(LL/N)_{true} - E\{LL/N\}] = -p/(2N). \tag{2.4.2}$$

Returning to L, instead of LL/N, the Akaike penalty for L is the exponential of this bias·N:

$$\text{penalty}_{Akaike} = \exp(\,-p/2\,). \tag{2.4.3}$$

Likelihood multiplied by this penalty function is called AIC. Let us remind that AIC was computed assuming N -> ∞, which is called asymptotical regime, and AIC is an asymptotical estimation. This is a significant deficiency, because often we would like to know parameters (and true number of models) from as little data as possible, for example, to have a useful financial prediction, we would like to obtain it from the minimal amount of past data. In addition, Akaike correction is shallow in shape, it does not produce a sharp maximum of L for the correct M value; so a lot of data N are needed to obtain correct estimation.

Because of the above criticism of AIC, often a different penalty function is used, related to Tikhonov regularization or Ridge regression

$$\text{penalty}_{\text{Tikhonov}} = \exp(-\gamma \sum_{m' \in M} |\mathbf{S}_m|^2). \tag{2.4.4}$$

Tikhonov or ridge penalty contains a coefficient, γ, which has to be selected empirically or heuristically; $|\mathbf{S}_m|^2$ is a sum of squares of all components of vector $\mathbf{S}_m$; so (2.4-3) penalizes for the sum of square of all parameters. A deficiency of this procedure is related to the fact that the value of a parameter depends on the scale used to measure it, and the value of a parameter depends on units of the scale. To improve efficiency of this procedure, one can carefully normalize all parameters, so that they would be non-dimensional, defined in terms of natural scales of the problem.

We would consider here one more penalty, based on "excessive" explanatory power of models. This basic idea has been originated by Vapnik (in the context of using no knowledge and no models). In our context, we interpret the Vapnik idea in the following way: the penalty should account for how flexible are models and conditional probabilities functional shapes. Functional shapes that can explain "all" existing knowledge cannot be predictive. In our interpretation, we relate this penalty function to the relative volume of data space explained by all the models and conditional similarities (except clutter model) vs. the total volume of the data space. The total volume of the data space is measured by

$$V_{\text{total}} = \text{volume}(\mathbf{X}) = \prod_{d} (X_{\max}-X_{\min})_d. \tag{2.4.5}$$

The product here is taken over all dimensions, d, of data **X**. The volume of the data space explained by a model $\mathbf{M}_m$ and conditional probability $\ell(n|m)$ is measured by a square root of the determinant of the covariance matrix of the shape $\ell(n|m)$, $(\det\mathbf{C}_m)^{1/2}$. For example of a Gaussian shape of $\ell(n|m)$, considered later, this equals the product of std over all data dimensions (in its principal coordinates). The total volume of the data space explained by all model is

$$V_{\text{models}} = \prod_{m} (\det\mathbf{C}_m)^{1/2}. \tag{2.4.6}$$

The product here is over all models except clutter. For non Gaussian shapes of $\ell(n|m)$, the covariance matrix of a conditional similarity can be expressed through model parameters. The corresponding Vapnik-Perlovsky penalty function we define as

$$\text{penalty}_{VP} = \exp(-\upsilon\ V_{models} / V_{total}). \quad (2.4.7)$$

Here, υ is a constant that should be determined heuristically.

Modifying similarity L by a penalty function leads to changes in eqs. (2.2-3) and (2.2-6):

$$d\mathbf{S}_m/dt = \sum_{n\in N} f(m|n)[\partial \ln\ell(n|m)/\partial\mathbf{M}_m]\partial\mathbf{M}_m/\partial\mathbf{S}_m + \partial(\ln \text{penalty})/\partial\mathbf{S}_m; \quad (2.4.8)$$

and

$$\mathbf{S}_m^{\ it+1} = \mathbf{S}_m^{\ it} + dt \cdot [\sum_{n\in N} f(m|n)[\partial \ln\ell(n|m)/\partial\mathbf{M}_m]\partial\mathbf{M}_m/\partial\mathbf{S}_m + \partial(\ln \text{penalty})/\partial\mathbf{S}_m]. \quad (2.4.9)$$

In the problem section we guide readers to derive these changes. In case of maximizing information and using equations from section 2.3, sums over n in the above equations are modified by weights abs($\mathbf{X}$(n)).

2.5 Convergence, Difficulties, and Solutions

On a first reading this section can be skipped. It is secondary to understanding of the main ideas and it might be too theoretical and too abstract before any experience with DL. Special situations will be discussed throughout the book, and it might be useful to return to this section from time to time, or when practically using DL for applications.

Many properties of the DL convergence can be understood by noticing that DL is a competitive procedure: due to the denominator in (2.2-1) models m "compete" with each other. Initially vague models and similarities with wrong parameter values, "grab" many wrong data points; eventually (sometimes after just few iterations) parameters tend to correct values, similarities tend to slimmer, crisper, concentrating around "their own" data points, the total similarity increases, and f(m|n) tends to 0 or 1.

DL is a convergent procedure, at the end of this chapter we reference literature where the convergence proof is given and also, in the problem section, we guide the readers to derive the proof. The convergence is guaranteed only to a local maximum. The DL process from vague to crisp smoothes out local maxima, while parameters are wrong and the global maximum is far away, still, this "smoothing out" the local maxima has its limit. Throughout the book we discuss how to achieve global convergence using simple rules. Here we briefly mention some heuristics used to avoid problems one may face when applying DL.

- An essential characteristic of DL process is evolution "from vague to crisp." Initial parameter values should be chosen so that std are large relatively to sensor errors, and are on the order of the entire range of data, (X_{max}-X_{min}), for each dimension.
- In the DL iterations, similarity (2.1-5) increases on each iteration; this can be used as a diagnostic tool to identify errors in a code.
- Initial parameter values should be chosen using available information, however approximate it might be; rarely do we face a problem with no prior knowledge whatsoever.
- Initial parameter values should be chosen so that std ranges of conditional similarities cover the expected range of solutions in space of signals, **X**(n).
- Initial models should not be identical in their parameters; identical models correspond to a local maximum, two models initialized with identical parameters will remain identical throughout iterations. If during the DL process two model become unreasonably close to each other, one should be switched into a dormant state.
- In a typical DL process, models and similarities overlapping in space of signals, **X**(n), tend to separate, each model "goes after" signals originating from a particular object or modeled process.
- *A fundamental suggestion, always have a "clutter" or "dust bin" model.* There are always many signals, which have nothing to do with objects or processes of your interest. It is necessary to get rid of these "clutter" signals (usually "noise" refers to errors in sensor measurements, whereas "clutter" refers to extraneous signals, unrelated to problems of interest). These clutter signals, if not discarded, can and will bias your solution or even completely destroy it (because in DL formulation, similar to probabilistic formulation, each data point has a total similarity (or probability) 1 of being associated with all models, according to (2.2-1). DL has a simple and powerful way to get rid of clutter, without thinking much of it. This is achieved by adding to the list of your models a "clutter," a "dust bin," or "everything else" model. The simplest such model, adequate in most cases is a constant throughout the entire signal space, throughout the book we assign this model a number m=1,

$$\ell(\mathrm{n}|1) = 1/\,\mathrm{volume}(\mathbf{X}). \tag{2.5.1}$$

The volume of **X** space, volume(**X**), eq. (2.3-5), is a normalization constant; practically, DL algorithms are not very sensitive to its exact value. The reason is that similarity (2.1-5) contains a product $r_1 \cdot \ell(\mathrm{n}|1)$. So this model has one unknown parameter, r_1, to be estimated from the data, and its estimated value would compensate for any inaccuracy in $\ell(\mathrm{n}|1)$. To summarize, a simple clutter-model contribution into total similarity could be

$$\text{Clutter contribution to likelihood, } r_1\ell(\mathrm{n}|1) = r_1/\mathrm{volume}(\mathbf{X}). \tag{2.5.2}$$

- A simple suggestion for choosing the number of models M: choose more models than you need. The final decision is made using a decision / detection criterion. It is used to decide which of the resulting models, upon convergence, are accepted as models of valid objects or processes, and which are rejected as a part of clutter. Such a detected criterion can be based on how many signals model m describes (that is, on the value of r_m); or it can be based on other meaningful expected properties of objects.

- *A fundamental rule:* there are local maxima corresponding to standard deviations (std) of conditional similarities -> 0. If std -> 0, ℓ(n|m) -> ∞, and correspondingly the total similarity, L -> ∞. This also causes numerical problems for the algorithm. In these cases often too few data points are assigned to models. If the number of parameters of model m is the same or larger that the number of data points associated with the model, the model exactly fits these data points, there is no error and std -> 0. In these cases f(n|m) -> 1 for fewer points n than the number of parameters in $\mathbf{M}_m$ and ℓ(n|m). To prevent these cases, a lower limit for std should be established (say equal to sensor errors, which are often known), and models achieving this limit should be reinitiated by significantly increasing its std (and possibly switching to a dormant status.
- DL equations (2.2-5, 2.2-6) and differential equation solvers require choosing a step, dt, for numerical solution. A too small dt might take needlessly long time to solve equations, which becomes important if models, $\mathbf{M}_m$, are obtained by numeric means and take long to solve (e.g., when solving inverse electromagnetic or hydrodynamic equations, $\mathbf{M}_m$ are "forward" solutions, which take long computations; or other similar cases). A too large dt results in erratic un-converging solutions. Few trials could be sufficient for choosing dt. If the overall computational time is not prohibitive, one may choose to err on a side of smaller dt.
- In DL, the balance between errors in parameter values and uncertainty of conditional probabilities is usually maintained automatically. However, sometimes, a conditional likelihood might jump to a "too" narrow range, too small std. A comprehensive DL algorithm corrects it through already discussed procedure: if std becomes too small, in few iterations it will rich the std lower limit (std of sensor error). This model should be eliminated, possibly replaced by a dormant model, etc. Sometimes one prefers to use a shortcut. If std is an explicit parameter of similarities (such as when using Gaussian functions), one can explicitly control std, using the following equation:

$$std = std_o \cdot \exp(-s\ t/T) + std_{sensor} \qquad (2.5.3)$$

Here std_o is the original large std, s is a heuristically selected parameter, t is an iteration "time", dt·iterations, T is the total time of all iterations, std_{sensor} is the final small std, determined by sensor errors. Using this shortcut requires some experience, s and T should be selected so that the convergence is completed by T, and the first item in (2.3-9) becomes significantly smaller than the second one. This approach can be combined with the standard estimation of covariance matrices, say stds on the next iteration can be chosen as the larger one of (2.5-3) and the one from standard estimation. This mixed approach preserves a useful DL property: the total similarity increases on each iteration.

- Sometimes convergence results are not as good as one expects from some extraneous information. Two suggestions are in order. First, possibly this additional extraneous information can be incorporated into the models to improve results. Second, there might not be sufficient information in available data, or may the algorithm has not reached the global maximum. Compare it to human vision system, which is very efficient, but not always perfect in a difficult condition. For example when skiing downhill a black dot in front of you could be some small insignificant object, or could be a big boulder on a next mountain. If one veers a bit left or right, the ambiguity is resolved. Similar with many real applications, if an object that you need to detect, is not detected from given data, the next moment new data comes, the next-next moment, more new data comes, and the needed object would be detected. If an algorithm works practically well enough, possibly one does not need to push it to the theoretical limit. (Even so we discuss theoretical limits in the book).
- The ultimate code debugging procedure. Simulate the data using the same models from the DL procedure, according to some selected parameter values. Initiate the DL process using exact known model parameter values. Add no clutter. In this case the DL procedure must converge after one iteration (to the correctly selected parameter values). If simulations must contain some random procedures, such as drawing data from probability distributions, select very small standard deviations. Allow for few iterations. If the solution deviates from the true parameter values more and more with iterations, the code contains errors.
- Numerical accuracy may become a problem, because conditional similarities $\ell(n|m)$, being exponential functions, or products of a large number of items, may become very small or very large (not likely). For preventing numerical problems, f(m|n) should be computed as follows,

  (1) Do not compute $\ell(n|m)$, but $\ell\ell(n|m) = \ln \ell(n|m)$, where logarithm expression should be computed analytically not to cause numerical difficulties;
  (2) Compute $\ell\ell\_max(n) = max_m[\ \ell\ell(n|m)\ ]$;
  (3) Normalize $\ell\ell\_norm(n|m) = \ell\ell(n|m) - \ell\ell\_max(n)$
  (4) Compute f(m|n) by using $\ell\ell\_norm(n|m)$; this procedure guarantees that the largest item in the denominator of f(m|n), and f(m|n) as well are computed with sufficient accuracy.

2.6 Problems

P2.6.1 Find on the web (e.g. on Wikipedia) literature on mixture models. Compare it to section 2.1 content. Write a short essay, 1/2 to 1 page, on similarities and differences.

P2.6.2 Compare solutions to mixture models in the literature to contents of section 2.2. Write a short essay, 1/2 to 1 page, on similarities and differences.

P2.6.3 Derive equations (2.2-1 through 2.2-4). Suggestions:

1) Consider LL = ln L
2) Maximize LL by gradient assent along parameters. For this, note that, you may start with any parameter values. Then, change LL through the "iteration time" t, or number of iterations, as follows d LL/d t = $(\partial LL/\partial \mathbf{S}_m)(\, d\mathbf{S}_m/dt\,)$. Note that if you choose $d\mathbf{S}_m/dt = \partial LL/\partial \mathbf{S}_m$, then d LL/d t = $(\partial LL/\partial \mathbf{S}_m)^2$ and therefore is positive, so with each step LL increases
3) To implement the above procedure, compute the gradient, $\partial LL/\partial \mathbf{S}_m$. When computing $\partial/\partial \mathbf{S}_m$ keep in mind that other models, $\mathbf{M}_{m'}$ for m' ≠ m, do not depend on $\mathbf{S}_m$. Use the following identities, d(ln(y))/dx = (1/y) dy/dx; d(Σy)/dx = Σ(dy/dx); and dy/dx = y d(ln(y))/dx.

P2.6.4 Derive equations (2.2-5 through 2.2-7).

P2.6.5 Derive equations (2.4-8 and 2.4-9). Follow P2.6.3

2.7 Literature for Further Reading

2.7.1 Section 2.1, Similarity Measure between Models and Data

Similarity between models and data: Perlovsky 2001; 2006a; Perlovsky & McManus, 1991;
Multiple hypothesis testing (MHT), Singer, Sea, & Housewright, 1974;
Dynamic Logic: Perlovsky 2001; 2006a,b; 2007b, 2009a, 2010c; Kovalerchuk & Perlovsky, 2008, 2009;
Reinforcement learning: Barto, Sutton, & Brouwer, 1981;
Statistical Learning Theory, SVM: Vapnik, 1998; Cherkassky & Mulier, 2007;

2.7.2 Section 2.2, DL Process from Vague to Crisp

Dynamic Logic: Perlovsky 2001; 2006a,b; 2007b, 2009a, 2010c;
Process from vague to crisp: Bar et al, 2006; Perlovsky 2009a, 2010c

2.7.3 Section 2.3, Mutual Information Similarity for Approximate Models

Mutual information similarity, Perlovsky 2001
Dynamic Logic: Perlovsky 2001; 2006a,b; 2007b, 2009a, 2010c;

2.7.4 Section 2.4, Number of Models

Akaike Information Criterion: Akaike, 1974; Perlovsky, 2001.
Tikhonov regularization and Ridge regression, see Wikipedia, http://en.wikipedia.org/wiki/Tikhonov_regularization
Statistical Learning Theory: Vapnik, 1998.

2.7.5 Section 2.5, Convergence, Difficulties, and Solutions

Competitive procedures: Grossberg, 1982; Carpenter & Grossberg, 1991.
Convergence of DL: Perlovsky, 2001.

Chapter 3
Classical Algorithms of Electrical Engineering and Signal Processing

DL leads to significant breakthrough improvements in solving classical algorithms. We consider detection; pattern recognition; clustering; joint detection, tracking, and association ("track before detect"); sensor fusion and association ("fuse before detect"); situational awareness; prediction and specifics of financial prediction. In all cases DL leads to practical improvements; we concentrate on complex aspects of problems, which have remained unsolvable for decades. Signals below clutter; situational awareness, as learning of what constitutes situations of interest vs. random collection of meaningless objects; in other words what context is, are examples of previously unsolvable problems.

3.1 Detection, Pattern Recognition and Data Mining

Detection of signals in clutter using models of signals and clutter, and recognition of patterns using models of patterns are mathematically equivalent. When models are exactly known, even exhaustive search is possible, its complexity is on the order of the total number of samples, N, times the number of samples per pattern, S, total $\sim N \cdot S$. If several patterns, M, should be found at once, the complexity is $\sim N \cdot S \cdot M$. This number might be large, but not incomputable. However, when models are characterized by parameters, which values are not known, the problem becomes excessively complex computationally. The only previously known general approach is called multiple hypothesis testing. It consists of two steps, first is a hypothesis which samples are associated with which model. Second, parameters of models are estimated, and a similarity measure between models and data is computed. These two steps are performed for all associations between models and data, and the highest similarity is selected. As discussed in previous chapters, this approach can not be practically realized, because the number of associations is excessively large $\sim M^N$; even for problems of moderate complexity this number is larger than the Universe; we call this difficulty Combinatorial Complexity, CC.

Here we consider several examples of solving this type of problems using DL. In the first example, we are looking for 'smile' and 'frown' patterns in clutter

L. Perlovsky et al.: Emotional Cognitive Neural Algorithms with Eng. Appl., SCI 371, pp. 31–80.
springerlink.com

shown in Fig.3.1A (section 3.1.3 below) without clutter, and in Fig.3.1B with clutter, as actually measured. Each pattern is characterized by a 3-parameter parabolic shape. The image size in this example is 100x100 points, and the true number of patterns is 3, which is not known. Therefore, at least 4 patterns should be fit to the data, to decide that 3 patterns fit best. Our standard estimation of multiple hypothesis testing complexity yields ~ 10^{60}. In this case another brute-force testing is possible, test all values of parameters possible within the grid. The number of parameters is 4x3=12, and within 100x100 grid all tests would take about 10^{32} to 10^{40} operations, still a prohibitive computational complexity.

To apply DL to this problem we need to develop parametric adaptive models of expected patterns, a uniform model for clutter, Gaussian blobs for highly-fuzzy, poorly resolved patterns, and parabolic models for 'smiles' and 'frowns'. In addition, we use this example to develop a slightly different version of similarity, which has a general applicability, when approximate models are used (as in this case).

3.1.1 Models for Detection, Example 1

The DL internal knowledge in this case is given by three types of parametric models: a uniform model for clutter, Gaussian blobs for vague, poorly resolved patterns, and parabolic models for 'smiles' and 'frowns.' The horizontal and vertical axes in images of Fig. 3.1, (x,y), in this example enumerate data samples, so we use a two-dimensional sample index, $\mathbf{n} = (n_x, n_y)$. Models and conditional similarities in DL are closely intertwined, so often it is convenient to discuss them jointly, and often we refer to conditional similarities as models and vice versa.

The clutter model is defined similarly to the discussion in the previous chapter. Scales along the horizontal and vertical axes (x,y) in images of Fig. 3.1 we arbitrary define to 1 unit, so that the conditional similarity is also defined to 1 unit,

$$\ell(\mathrm{n}|1) = 1/\mathrm{volume(data\ space)} = 1. \qquad (3.1.1)$$

The only parameter of the clutter model is its rate, r_1.

Gaussian blob models are defined by their parameters: rates, r_m; central locations, $\mathbf{n}_m$, which are two-dimensional positions along (x,y), $\mathbf{n}_m = (n_{mx}, n_{my})$; standard deviations in x and y, σ; and by their vertical shapes, modeled by Gaussians, which are used as conditional similarities,

$$\ell(\mathrm{n|m, Gaussian}) = \mathrm{G}(\mathbf{n}|\mathbf{n}_m, \sigma) = (1/2\pi\sigma) \bullet \exp[-(\mathbf{X}(\mathrm{n}) - \mathbf{n}_m)^2 / 2\sigma^2];. \qquad (3.1.2)$$

These are bell-shapes above the **X**-plane with the center at $\mathbf{X} = \mathbf{n}_m$, and a radial size ~ σ.

Each parabolic model is defined by its parabolic shape in (x,y), and its vertical shape is modeled approximately as a superposition of K Gaussian components located along the parabolic shape in (x,y); correspondingly, the rate is not a single variable, but each of the K components has its own rate, r_{km}. This is not the only way to model 3-dimensional shapes, but it is a convenient, relatively parsimonious, and pretty general one, as will be seen from subsequent examples. The parabolic shape in (x,y) is given by central locations of K components,

$$\mathbf{n}_{mk} = (n_{mkx}, n_{mky}) = (k_{mx}, k_{my}) + (k, a_m k^2); \tag{3.1.3}$$

Here (k_{mx}, k_{my}) is a center location of the parabolic model, k is an index of components, k = -K/2,… K/2. The number of components, K, should be chosen to satisfy two contradictory conditions: the larger is K, the more smooth is the modeled shape, closer to the shape of data, but then more computations are required; somewhat arbitrary we have chosen K = 10 (if there is no information for choosing K, it could be estimated along with other parameters; and it could be different for different models). Parameter a_m determines the curvature of the parabolic shape. The amplitude-shape of a model ("above" the image plane), similar to Gaussian models in (3.1-2), is included into the conditional similarity, which is taken as a superposition of K Gaussian components,

$$l(n|m, \text{parabolic}) = \sum_{k \in K} r_{km}\, G(\mathbf{n}|(\mathbf{n}_{mk}, \sigma). \tag{3.1.4}$$

Due to weights, abs($\mathbf{X}$(n)), in the DL equations, centers of conditional similarities, $\mathbf{n}_{mk}$, are spread along the maximal signal values so that K Gaussian components uniformly model every "smile" and "frown" in Fig. 3.1A, as discussed in the next section.

To compute r_{km}, we use iterative equations for f(m|n) and r_m given in section 2.3:

$$\ell(n|k,m) = G(\mathbf{n}|(\mathbf{n}_{mk}, \sigma). \tag{3.1.5}$$

$$f(k,m|n) = r_{km}\, \ell(n|k,m) / \sum_{k',m'} r_{k'm'}\, \ell(n|\, k',m'). \tag{3.1.6}$$

$$r_{k,m} = 1/N \sum_{n \in N} \text{abs}(\mathbf{X}(n))\, f(k,m|n). \tag{3.1.7}$$

As always, solving iteration equations, on each iteration we compute r.h.s using parameters from the previous iteration, and l.h.s. contains the new, current values, used to compute new parameters.

3.1.2 Detection, Example 1

In this example, we are looking for 'smile' and 'frown' patterns in clutter shown in Fig.3.1.1A without clutter, and in Fig.3.1.1B with clutter, as actually measured. We use models described in previous section: clutter, circular Gaussian blobs, and parabolic shapes (although expected 'smiles' and 'frowns,' Fig.3.1.1A, are not exactly parabolic).

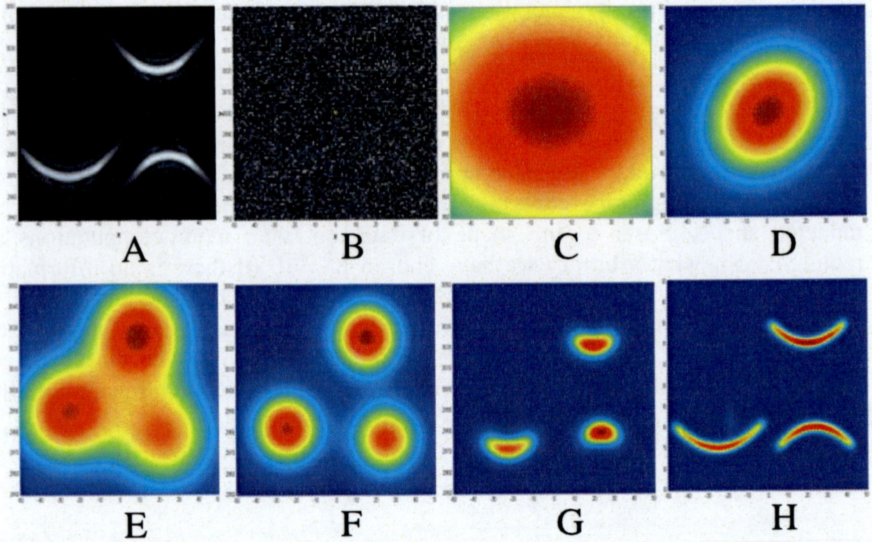

Fig. 3.1.1 Finding 'smile' and 'frown' patterns in noise, an example of dynamic logic "from vague-to-crisp" process: (A) true 'smile' and 'frown' patterns are shown without clutter; (B) actual image available for recognition (signal is below clutter, signal-to-clutter ratio is about 1/3); (C) an initial fuzzy blob-model, the vagueness corresponds to uncertainty of knowledge; (D) through (H) show improved models at various iteration stages (total of 21 iterations). Between stages (D) and (E) the algorithm tried to fit the data with more than one model and decided, that it needs three blob-models to 'understand' the content of the data. Until stage (G) the algorithm 'thought' in terms of simple blob models, at (G) and beyond, the algorithm decided that it needs more complex parabolic models to describe the data. Initial models contain low-spatial frequencies compared to the final one. Iterations stopped at (H), when similarity (2.3.1) stopped increasing.

The initial state of the DL process has an active clutter model (not shown in images) with r_1 initiated to the total signal amplitude, one active Gaussian blob model, initiated with standard deviations on the order of the total image and $r_2 = 0.1r_1$ (an arbitrary choice), shown in Fig.3.1.1C. There are also 2 inactive models, Gaussian blob and "smile" (not shown). The first iteration in Fig.3.1.1D is already noncircular, indicating that two Gaussian blobs are active. In iteration 5, Fig.3.1.1E, three Gaussian blobs are active. In iteration 11, Fig.3.1.1F, still three Gaussian blobs are active, while uncertainty is reduced. When models come close to the true shape, iteration 17, Fig.3.1.1.G, there is sufficient sensitivity to determine that parabolic shapes better match signals, three parabolic shapes are activated, whereas Gaussian blobs become inactive. At iteration 21, Fig.3.1.1H, iterations stop, because similarity (2.3.1) stopped increasing with iterations (a threshold used to evaluate similarity changes is somewhat arbitrary; it depends on a particular problem and its selection requires few trials; selecting a very small threshold will not significantly increase the number of iterations).

The number of computer operations in this example was about 10^9. Thus, a problem that was not solvable due to CC becomes solvable using DL.

To summarize this example, during an adaptation-learning DL-process, initial vague-fuzzy and uncertain models are associated with structures in the input signals, and vague models become more definite and crisp with successive iterations. The type and shape of models are selected so that the internal representation within the system is similar to input signals for some unknown values of parameters. In this sense, the DL models, after adaptation, approximately represent structure-objects in the signals; images in Fig.3.1.1A are only approximately parabolic. In the image available for recognition, Fig.3.1.1B, signal is below clutter, signal-to-clutter ratio is about 0.3. This is a significant improvement over other state-of-the-art practically working algorithms; a standard required signal-to-clutter ratio is more than 30 (references, as always, are discussed at the end of the chapter). The achieved improvement is about 100 times.

3.1.3 Detection of Moving Objects, Example 2

In this example the knowledge about expected objects is that, in addition to clutter, there might be elongated objects, moving along unknown straight paths and rotating with unknown speed; exact shape of objects is unknown, strength of signals is expected to be below clutter and a desirable signal-to-clutter ratio could be as low as 1/5 (a bit lower than in the previous example). A sequence of 25 images, 256 x 256 pixels each, is available for processing. Problems of such complexity have not been previously considered.

We use the same similarity measure and similar models for clutter and Gaussian blobs to those discussed in the previous example. The purpose for using Gaussian blobs is two-fold. First, at the beginning of the DL-process, when models significantly differ from structures in the signal, Gaussian blobs are adequate and faster to compute. Second, clutter might be non-uniform, objects different from those of interest might be present, so they are to be captured by Gaussian blobs (we remind a fundamental property of DL: since every data point is actually present with 100% probability, its 100% existence has to be "explained" by some models). A model for moving and rotating objects is given by

$$\ell(\mathbf{X}(n)|m = \text{moving, rotating}) = (1/2\pi)\det\mathbf{C}_m^{-1/2} \bullet \exp[-(\mathbf{X}(n) - \mathbf{M}_m)^T \mathbf{C}_m^{-1} (\mathbf{X}(n) - \mathbf{M}_m)/2]; \quad (3.1.8)$$

$$\mathbf{M}_m = (X_m + T\bullet V_{mx}, Y_m + T\bullet V_{my}); \quad (3.1.9)$$

$$\mathbf{C}_m = \text{diag}(\, C1_m + C2_m \cos(T\bullet\omega_m), C1_m + C2_m \sin(T\bullet\omega_m)\,). \quad (3.1.10)$$

Here, $\mathbf{M}_m$ is a center of object m moving with velocity $\mathbf{V}_m = (V_{mx}, V_{my})$; $\mathbf{C}_m$ is a diagonal covariance determining a rotating elongated shape; according to existing knowledge, somewhat approximately we set $C2_m = 100\ C1_m$; T is a time of actual object motion and rotation; ω_m is a frequency of the object rotation. Parameters of these models included in $\mathbf{S}_m$ are $(X_m, Y_m, V_{mx}, V_{my}, C1_m, \omega_m, r_m)$. Note we use the same letter T in a superscript in (3.1-8) to denote a transposed vector (this has nothing to do with time, and we hope there could be no confusion); whereas $\mathbf{X}(n)$ and $\mathbf{M}_m$ are column-vectors, $(\mathbf{X}(n) - \mathbf{M}_m)^T$ are row-vectors. MATLAB and other higher languages operating with matrixes take care of this automatically; some lower-level languages use functions for vector-matrix operations, and others require explicit specification of indices, for example

$$[-(\mathbf{X}(n) - \mathbf{M}_m)^T \mathbf{C}_m^{-1} (\mathbf{X}(n) - \mathbf{M}_m)] = \sum_{i,j} (X_i(n) - M_{m,i}) (\mathbf{C}_m^{-1})_{i,j}(X_j(n) - M_{m,j}). \tag{3.1.11}$$

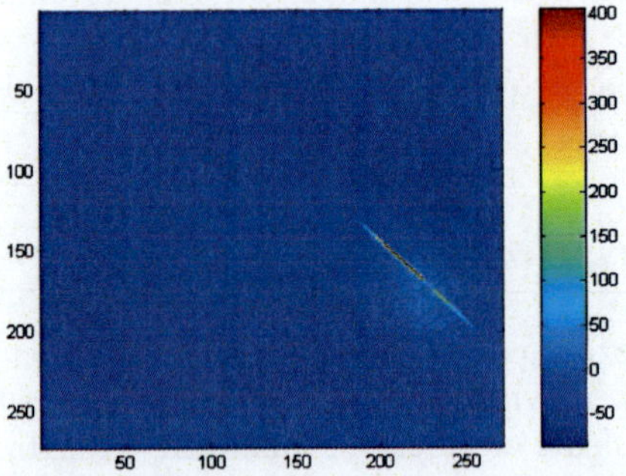

Fig. 3.1.2a One frame S/C = 300.

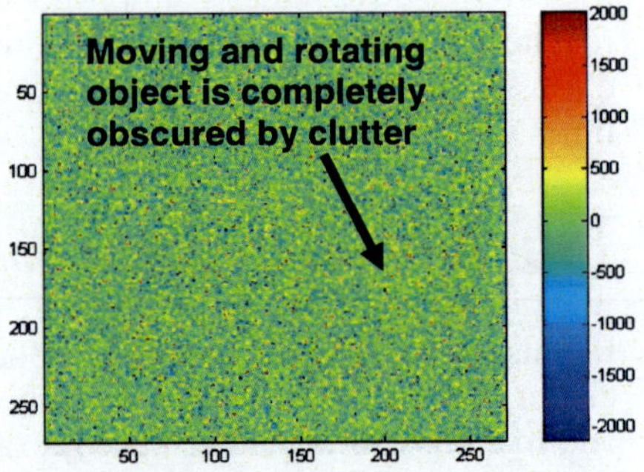

Fig. 3.1.2b One frame S/C = 0.2.

Fig. 3.1.2c DL processing of Fig. 2b data, iteration 10.

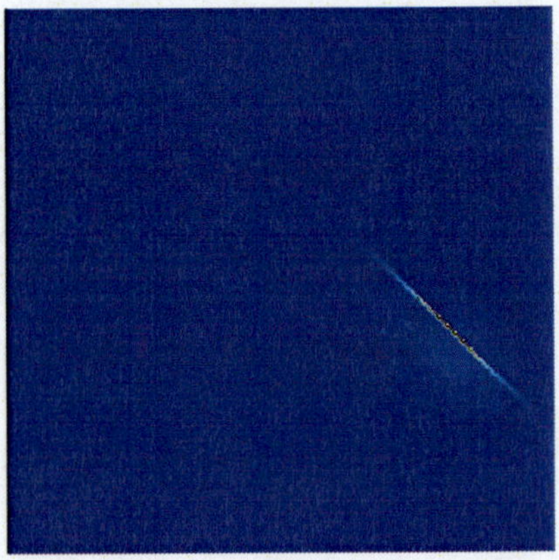

Fig. 3.1.2d DL convergence results, iteration 600.

Parameters, $\mathbf{S}_m = (X_m, Y_m, V_{mx}, V_{my}, C1_m, \omega_m, r_m)$ in this case are estimated using equations (2.3-2 through 2.3-4), with derivatives computed numerically. Whereas the DL algorithm does not know the number of the moving and rotating objects and is set to detect several of them, the example illustrated in Fig.3.1.2 has just one such object, and it shows one frame of this moving and rotating object along with the models in this frame. Fig.3.1.2a shows one representative frame of a moving and rotating object measured with low clutter, so that signal-to-clutter ratio (S/C) is about 300, which have been previously considered necessary for a reliable object detection; on the right is a signal strength-to-color mapping bar in arbitrary units. Fig.3.1.2b shows similar image at realistic signal-to-clutter ratio of interest, S/C = 0.2.

Fig.3.1.2c shows a DL iteration 10; there are 26 activated blob-models with relatively low strength (activation threshold was set at r_m = 0.0001 of the total signal strength) and 1 vague activated moving and rotated elongated object model (uniform clutter model is not shown). Intermediate frames with motion and rotation of the object are not shown. Fig.3.1.2d shows a DL convergence results at iteration 600. The total number of operations is about 10^{12}. (Possibly the number of iterations could be significantly reduced, no efforts were devoted to this). The model of the elongated object as well as surrounding clutter blobs are estimated closely to the image acquired at high image-to-clutter ratio, Fig.3.1.2a, which previously was not considered possible. The S/C improvement is about 1,500 (to emphasize this point, 150,000% improvement).

3.2 Clustering

3.2.1 The Problem and DL Equations

In many applications including machine learning, data mining, bioinformatics, financial prediction, pattern recognition, image analysis, etc., it is necessary to find "natural" grouping or clustering of data. This area of statistical data analysis is called clustering or cluster analysis. We took "natural" in quotes, because what is natural depends on existing knowledge for a particular problem. From this point of view, DL is a powerful clustering technique, because it enables an easy utilization of diverse and complex existing knowledge. Detection problems in section 3.1 can be considered as this kind of clustering with highly specific knowledge. Another point of view on clustering is that it is often used as an exploratory investigation, when little specific knowledge about the problem exists. In such cases one tries to gather as much data as possible, and to find groups or clusters using as little knowledge as is available. Diverse knowledge about every data point is usually represented as a multi-dimensional vector, $\mathbf{X}(n) = (x_1(n)\ldots x_d(n))$. These d components are often called dimensions, features, or attributes. The d-dimensional space of $\{\mathbf{X}(n)\}$ is addressed as statistical, clustering, or feature space. Below, this section considers a situation of multi-dimensional clustering with no prior knowledge.

Among difficulties of clustering are (1) finding reliable clustering from few data points; this is especially important, when data points are limited; for example in financial predictions one would like to use only the most recent data; in bioinformatics sometimes every data point is a result of an expensive experiment, so data points could be expensive to measure; (2) finding reliable clustering in high dimensional clustering space, in other words, when every data point has many attributes or features (since the 1960s this difficulty is called "curse of dimensionality"); (3) avoiding false clusters (local maxima of a measure used for clustering).

DL addresses difficulty (1) by using parametric cluster shapes with few parameters (the fewer parameters, the fewer data points are needed to estimate them); (2) using few parameters in high dimensional data spaces as discussed below; (3) using the DL-process from vague to crisp, which smoothes, to some extent, local maxima of the similarity.

DL for clustering below uses a similarity measure and DL equations from section 2.2 (although 2.3 equations could be used as well). Clutter model usually is not needed (the reason is that clutter will be assigned to its own cluster); unless a lot of clutter is expected (data points of no interest), while of interest are few important clusters. In DL clusters are usually described by Gaussian conditional similarities (unless, specific information is available for using other shapes) similar to (3.1-8)

$$\ell(\mathbf{X}(n)|m) = (1/2\pi)^{d/2} \det\mathbf{C}_m^{-1/2} \bullet \exp[-(\mathbf{X}(n) - \mathbf{M}_m)^T \mathbf{C}_m^{-1} (\mathbf{X}(n) - \mathbf{M}_m)/2]. \quad (3.2.1)$$

Here d is dimensionality; data $\mathbf{X}(n)$ and models $\mathbf{M}_m$ are d-dimensional vectors; $\mathbf{C}_m^{-1}$ is a d-dimensional inverse covariance matrix.

Clustering with Gaussian conditional similarities is known as Gaussian mixture model (GMM). Mixture refers to sums in (2.1-2, 2.1-5), *all DL models are mixture models;* historically Gaussian mixtures were widely studied (yet, before developing DL they were often considered too complicated for practical use).

To keep few parameters, covariance matrices can be limited to diagonal shape. In this case the number of parameters per cluster is 2d+1 (d for the model-mean $\mathbf{M}_m = (M_{m,1\ldots}\ M_{m,d})$; d for the covariance diagonal values $\mathrm{diag}(\mathbf{C}_m) = (C_{m,1\ldots}\ C_{m,d})$; these are squares of standard deviations, $C_{m,i} = \sigma^2_{m,i}$; and 1 parameter for r_m). For diagonal covariances, inverses are simply related to standard deviations, $(\mathbf{C}^{-1})_{m,i} = \sigma^{-2}_{m,i}$. One can consider $\sigma^{-2}_{m,i}$ as parameters and estimate them directly. For diagonal covariances, determinants are simply products of the diagonal elements:

$$\det\mathbf{C}_m^{-1/2} = \sigma^{-1}_{m,1} \ldots \bullet\sigma^{-1}_{m,i} \ldots \bullet\sigma^{-1}_{m,d}. \quad (3.2.2)$$

Limiting covariance matrices to diagonal shapes ignores possible correlations among dimensions, which could be different among clusters, but it significantly reduces the number of parameters. To further reduce the number of parameters, one can use the same standard deviations for all dimensions and all clusters, $\sigma^{-2}_{m,i} = \sigma^{-2}$. If a lot of data are available, full covariance matrices can be used.

In this case it is interesting to look at Choleski decompositions of covariance matrixes, which are even more convenient to apply to the inverse covariances,

$$\mathbf{C}^{-1}{}_{m} = \mathbf{H}_{m}{}^{T}\,\mathbf{H}_{m}. \tag{3.2.3}$$

Choleski factors defined this way $\mathbf{H}_m$, are higher triangular matrices. A determinant of a triangular matrix is just a product of its diagonal elements,

$$\det\mathbf{C}_m{}^{-1/2} = \det\mathbf{H}_m = \det\mathbf{H}_m{}^{T} = H_{m,1} \ldots \bullet H_{m,i} \ldots \bullet H_{m,d}. \tag{3.2.4}$$

In terms of Choleski factors, the exponent in (3.1-1) simplifies, and Gaussian similarity (3.1-1) can be re-written as

$$\ell(\mathbf{X}(n)|m) = (1/2\pi)^{d/2}\det\mathbf{H}_m \bullet \exp[-\,(\,(\mathbf{X}(n) - \mathbf{M}_m)^T \bullet \mathbf{H}_m{}^T\,)^2/2]. \tag{3.2.5}$$

In this case one does not need to estimate covariances, instead, parameters are the Choleski factors, $\mathbf{H}_m$ (or $H_{m,i,j}$) for $i \le j$).

Instead of eq.(2.2-6), simpler equations can be derived for estimating parameters of Gaussian similarities (as usual, in the problem section we guide readers to derive these equations). Beginning from some initial values, on every iteration we compute f(m|n) and r_m using standard equations

$$f(m|n) = r_m\,\ell(n|m) \,/ \sum_{m' \in M} r_{m'}\,\ell(n|m'). \tag{3.2.6}$$

$$r_m = (1/N) \sum_{n \in N} f(m|n). \tag{3.2.7}$$

Simpler equations for parameters of Gaussian similarities are given by

$$\mathbf{M}_m = (1/N) \sum_{n \in N} f(m|n)\,\mathbf{X}(n). \tag{3.2.8}$$

$$\mathbf{C}_m = (1/N) \sum_{n \in N} f(m|n)\,(\mathbf{X}(n) - \mathbf{M}_m)\,(\mathbf{X}(n) - \mathbf{M}_m)^T. \tag{3.2.9}$$

When standard deviations are all equal σ,

$$\sigma^2 = 1/(N \bullet M \bullet d) \sum_{n,m} f(m|n)\,(\mathbf{X}(n) - \mathbf{M}_m)^2. \tag{3.2.10}$$

This leads to a more accurate estimation, than for full covariance matrices. (Of course, (3.2-10) is equivalent to averaging (3.2-9) over all models m, and dimensions d.

Despite of simplified expression for $\ell(\mathbf{X}(n)|m)$ in terms of $\mathbf{H}_m$, (3.2-5), estimation of $\mathbf{H}_m$, still requires first, to estimate $\mathbf{C}_m$, eq.(3.2-9), then invert it to receive $\mathbf{C}^{-1}{}_m$, then decompose it to Choleski factors. Therefore, one would rather use eqs.(2.2-5 through 2.2-7) for estimating directly Choleski factors of the inverse covariance (as well as all other parameters).

Equations (3.2-6) through (3.2-10) therefore are not "simpler" to solve than the general DL equations in chapter 2. One may argue that chapter 2 equations are "easier," you code them once, and then apply to any problem in this book. And one doesn't need to invert covariance matrices, Choleski factors of inverse covariances can be used instead, which leads to a great speed-up for multi-dimensional clustering. And we advocate this approach. Nevertheless, eqs.(3.2-6) through (3.2-10) have an advantage of intuitive interpretation: all parameters are estimated similarly to a most standard problem of estimating the mean and covariance of a data set, if no clustering is required. The only difference here is that first, it is an iterative process and second, on each iteration, averaging is weighted with f(m|n) weights, which assign every data element (n) to its cluster (m) with a proper weight. If clusters are not overlapping, but well separated, f(m|n) converge to 0 or 1, and in the result, parameters of each cluster are estimated independently from each other (as in the standard statistical estimation with only 1 group of data). Equations (3.2-6) through (3.2-10) are useful for low-dimensional cases, when inverting matrices does not take much of computer time, and it might be especially instructive at the learning stage.

3.2.2 DL Clustering, Example 1

The data for this example is generated using two Gaussian 2-D distributions, illustrated in Fig. 3.2.1. The dynamic logic algorithm given by (3.2-6-8) is applied to the data and converges within 20 iterations.

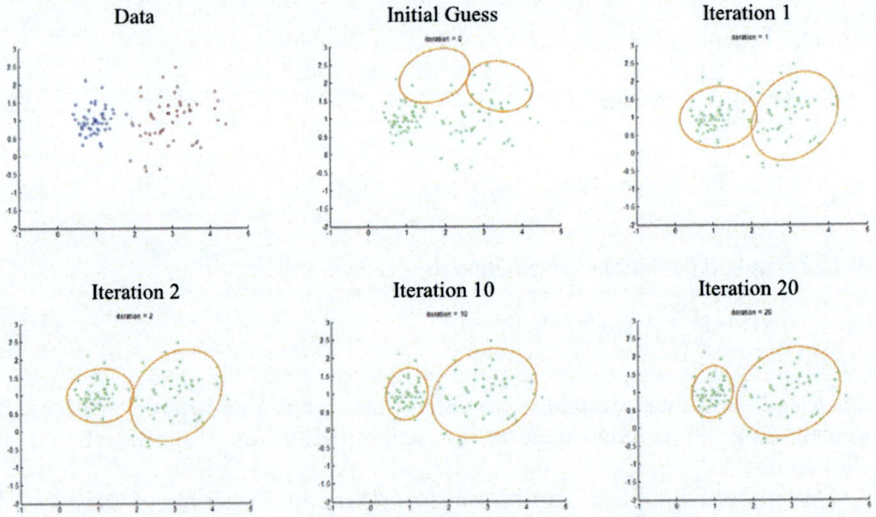

Fig. 3.2.1 Clustering in 2 dimensions. The data is generated using 2-dimensional Gaussian distributions. Class one has mean [1,1] and variance 0.1. Class two has mean [3, 1] and variance 0.4.

Example 2

This example uses the time series representing 8 normalized daily financial indicators over approximately one year time period. The data is shown in Fig. 3.2.2.

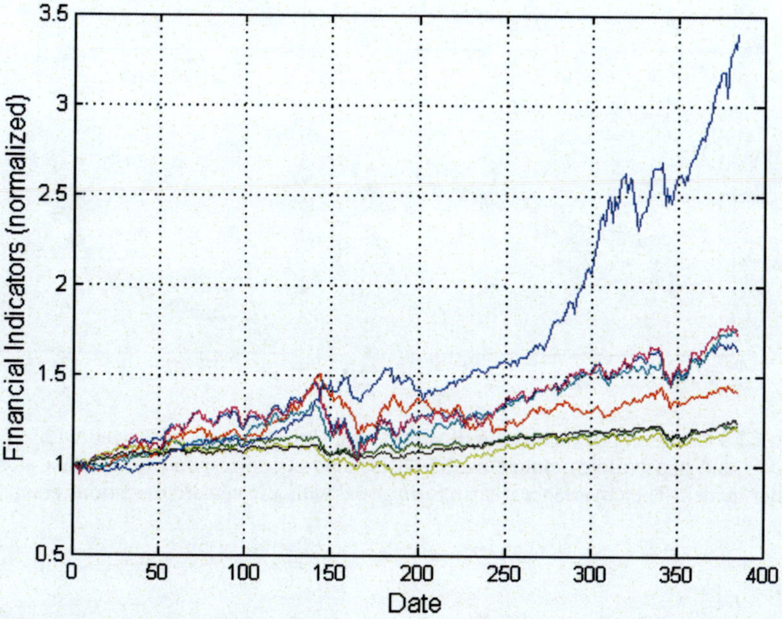

Fig. 3.2.2 Normalized financial indicators.

Clustering technique analogous to that of Example 1 is used in this example to identify groupings within the data. Since the data is 8-dimansional it is only possible to visualize two-dimensional projection of the data. In important difference between this example and the previous one is the presence of clutter model, described by uniform probability density.. The clutter model captures data points that do not belong to any cluster resulting is more compact clusters.

To illustrate various degrees of model complexity, consider three different cases with progressively more parameters involved. Figure 3.2.3. illustrates the results of clustering with diagonal covariance matrix and equal standard deviations for all dimensions. Such simplified model does not fully capture the shape of clusters and many data points are assigned to the clutter model. The condition of equal standard deviations is relaxed in Fig. 3.2.4. Finally, the full covariance matrix is used in Fig. 3.2.5. The full covariance allows better fit to the data and fewer points assigned to clutter.

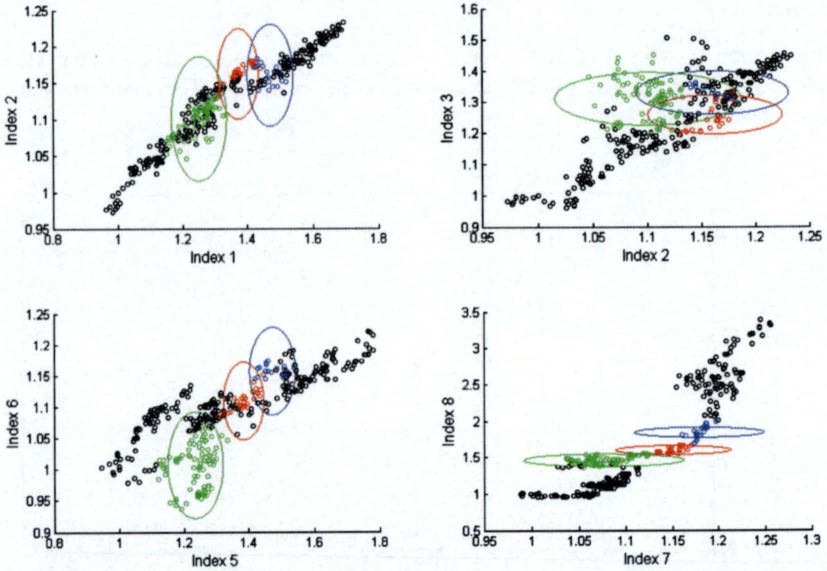

Fig. 3.2.3 Clustering of financial indicators. The algorithm was initialized with 3 Gaussian models and one uniform clutter model. Black points correspond to the data assigned to clutter model. The covariance matrix is diagonal with all standard deviations equal.

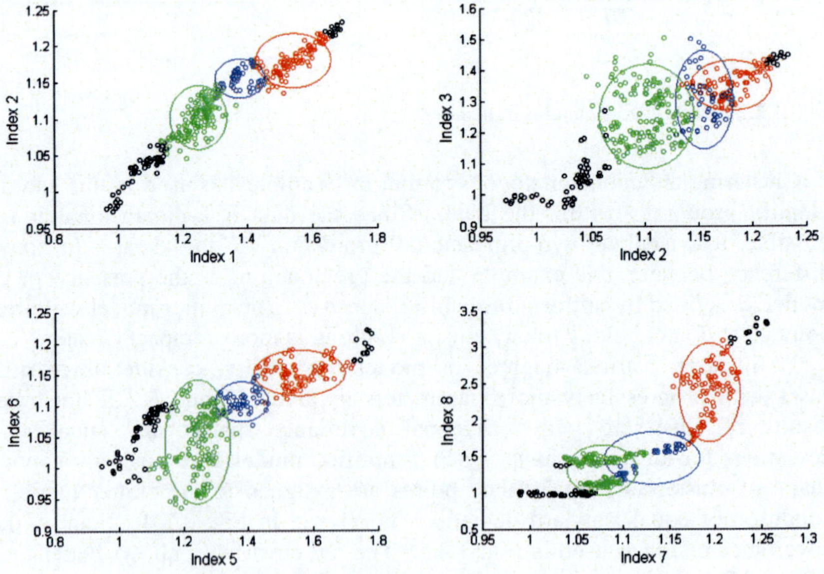

Fig. 3.2.4 Clustering of financial indicators. The algorithm was initialized with 3 Gaussian models and one uniform clutter model. Black points correspond to the data assigned to clutter model. The covariance matrix is diagonal.

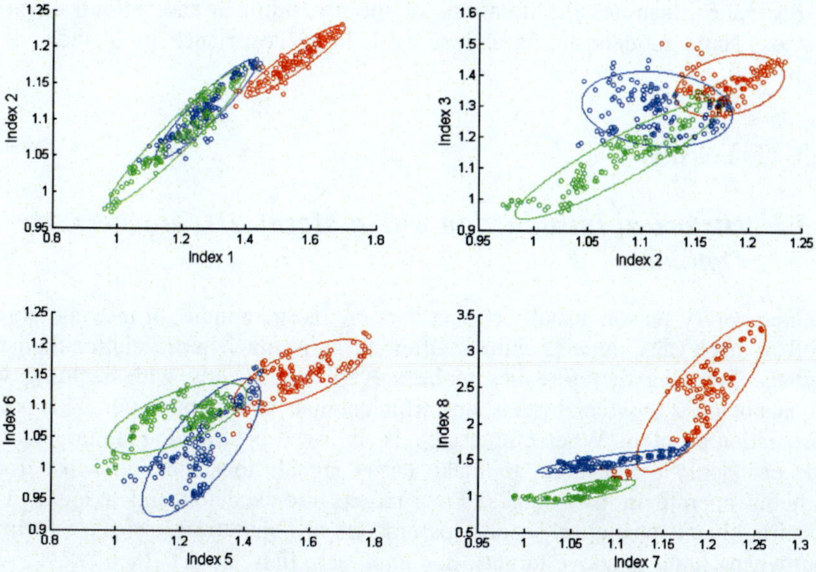

Fig. 3.2.5 Clustering of financial indicators. The algorithm was initialized with 3 Gaussian models and one uniform clutter model. Black points correspond to the data assigned to clutter model. The full covariance matrix is used.

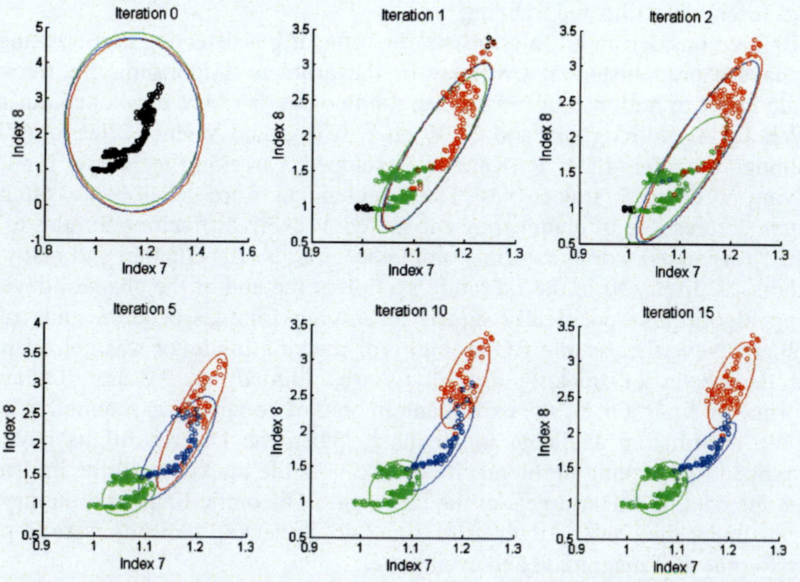

Fig. 3.2.6 Iterations of the dynamic logic algorithm for clustering of financial indicators are shown for indicators 7 and 8. The algorithm was initialized with 3 Gaussian models and one uniform clutter model. Black points correspond to the data assigned to clutter model. The full covariance matrix was used.

Fig 3.2.6 illustrates the iterations of the algorithm in case of full covariance matrix. Three models are initialized with large covariance to include all data points.

3.3 Tracking

3.3.1 Historical Introduction with a Moral: DL Trackers Are Optimal

Contemporary sensors usually collect data on a large number of moving targets as well as irrelevant signals, clutter. Often there is much more clutter than target signals. Tracking therefore has to be solved concurrently with deciding which signals belongs to which targets, and which signals belong to clutter. This is called association problem. When clutter signals are lower in strength than target signals, one can apply a threshold, to make target signals to stand out. Most tracking systems operate in two stages. First targets are detected and associated with specific objects or clutter, second, parameters of object trajectories are estimated. But when clutter is above targets, this approach, first "detect" then "track," is not applicable. When the only difference between targets and clutter is that targets are moving along specific trajectories, association and tracking have to be solved concurrently, so that targets signals can be associated through several scans. Sometimes it is called "track before detect." A more accurate is to call it "concurrent detection and tracking."

Before considering DL algorithms for "concurrent detection and tracking," let us make a brief historical overview of the problem. Algorithms for tracking a single target in presence of radar noise (and no clutter) were developed during the WWII by A. Kolmogorov and N. Wiener. It is called Wiener filter, or Wiener-Kolmogorov filter. Later R. Kalman developed a tracking technique for targets moving on complex trajectories. These techniques were developed assuming no clutter. In presence of clutter they encountered severe difficulties, similar to those already discussed, combinatorial complexity, CC. Y. Bar-Shalom and many other authors, as discussed in the literature section at the end of the chapter, developed many algorithms specifically aimed at solving joint association and tracking problem. Nevertheless, the CC problem for tracking in clutter was not overcome, and the power of tracking algorithms was limited not by any fundamental information limit, but by the exorbitant amount of computation required to extract all the information available in the data. When the DL algorithms have been developed for tracking problems, it become possible to extract all the information from the data, to track targets at the information-theoretic limit, and literary track targets under the clutter. In terms of signal-to-clutter ratio, results were improved by two orders of magnitude and even better.

Understanding of this historical development is important, because, for some mysterious reason (possibly like any other fashion) every several years, a new tracking algorithm gains popularity, while remaining limited by CC. Currently, particle filter algorithms gain popularity, while remaining bound by the old limits of CC and under-perform DL trackers by orders of magnitude. Therefore we would like to emphasize once more that DL algorithms for tracking in clutter perform near the information theoretic limits. This mathematically best possible performance can not be overcome by a better or more fashionable mouth-trap. DL algorithms are also simple in implementation; therefore there is no need to look for other solutions; it is a mathematical fact that the performance of DL algorithms achieves the information-theoretic limit, and cannot be improved by any other algorithm.

3.3.2 DL Equations

Many radar systems measure positions of targets in two directions (x, y) called range and cross-range, a Doppler velocity in range direction, D, and also an amplitude or strength of the signal, a. Correspondingly, we denote signals $\mathbf{X}(n) = (x_n, y_n, a_n, D_n)$. To be specific we discuss so called ground moving target indicator (GMTI) radar, which measures signals reflected mostly by the ground and objects on the ground. Accuracy of measurements in range and Doppler are usually much higher than in cross-range and amplitude. Both similarity measures, from sections 2.2 or 2.3 can be used for tracking; as usual, section 2.2 equations are appropriate when signals passed through a threshold, and one deals with individual signals; when $\mathbf{X}(n)$ form a continuous image in (x, y), equations from section 2.3 should be used. We consider the case of isolated measurements in (x, y), and consider tracking short track segments, tracklets, along which velocities can be considered constant $\mathbf{V}_m = (V_{mx}\ ,\ V_{my})$. Correspondingly, the complete model is

$$\mathbf{M}_m(\mathbf{S}_m,n) = (X0_m+V_{mx}T_n,\ Y0_m+V_{my}T_n,\ a_m,\ D_m). \tag{3.3.1}$$

Here parameters of the model, $\mathbf{S}_m = (X0_m, Y0_m, V_{mx}, V_{my}, a_m, D_m)$; $(X0_m, Y0_m)$ model an original position, (V_{mx}, V_{my}) model velocity, (a_m, D_m) model amplitude and Doppler; T_n, is the known time counted from the first scan. Also,

$$V_{mx} = D_m. \tag{3.3.2}$$

Practically, there is no need to treat (3.3-2) as a constrain, additional to (3.3-1), one just can use V_{mx} instead of D_m or vice versa. The unknown parameters also include r_m, parameters of conditional similarities, such as standard deviations or covariances, and the total number of track-models. Conditional similarities for clutter we define as uniform, according to (2.4-1)

$$\ell(n|1) = 1/\ \mathrm{volume}(\mathbf{X}). \tag{3.3.3}$$

Conditional similarities of tracks are defined as Gaussian. Although radar signals a_m are not likely to follow Gaussian distributions, in our practical cases this approximate treatment has been sufficient,

$$\ell(n|m) = (2\pi)^{-2}(\det \mathbf{C}_m)^{-0.5} \cdot \exp[-(\mathbf{X}(n) - \mathbf{M}_m)^T \mathbf{C}_m^{-1} (\mathbf{X}(n) - \mathbf{M}_m)/2]. \tag{3.3.4}$$

We use diagonal covariance matrixes $\mathbf{C}_m = \mathrm{diag}(\sigma_{xm}^2, \sigma_{ym}^2, \sigma_{am}^2, \sigma_{Dm}^2)$; and, $\sigma_{xm}^2 = \sigma_{Dm}^2$.

The DL iterative equations for estimating these parameters in this case are earthier to write with the following notations

$$<\ldots>_m = \sum_{n \in N} f(m|n)\,(\ldots)_n. \qquad (3.3.5)$$

Then parameter estimation equations, at each iteration, are computed as

$$r_m = <1>_m / N. \qquad (3.3.6)$$

$$a_m = <a_n>_m. \qquad (3.3.7)$$

$$Y0_m <1>_m + V_{ym} <T_n>_m = <Y_n>_m,$$

$$Y0_m <T_n>_m + V_{ym} <T_n^2>_m = <Y_n T_n>_m. \qquad (3.3.8)$$

$$X0_m <1>_m + V_{xm} <T_n>_m = <X_n>_m,$$

$$X0_m <T_n>_m + V_{xm} (<T_n^2>_m + c<1>) = <X_n T_n>_m + c <D_n>_m. \qquad (3.3.9)$$

Here, $c = \sigma_{xm}^2 / \sigma_{Dm}^2$. For the unknown parameters, $Y0_m$ and V_{ym}, eqs.(3.3-8) is a two-dimensional linear system of equations; similarly eq.(3.3-9) is a two-dimensional linear system of equations for $X0_m$ and V_{xm}; these equations should be solved at every iteration, which is of course easy to code or can be done by standard linear equation solvers. Standard deviations for each parameter s are estimated, as follows:

$$\sigma_{ms}^2 = <(X_s(n) - M_{ms}(n))^2>_m \qquad (3.3.10)$$

One can question if eqs.(3.3-6) through (3.3-10) give any advantage compared to eqs.(2.2-2) through (2.2-7). We repeat that the general equations from section 2 are easier to use, especially if they have been already coded and the only required modification are models, (3.3-1, 3.3-2), which of course could be easily combined in a single equation.

We would suggest that for tracking and for some other applications the criterion for stopping iterations could be changed from "global" to "local." That is, one can stop iterations for each track independently, either, when parameters of this track stop changing significantly, or if a local similarity for this track exceeds a predetermined threshold. For this purpose we define a local log-similarity for track m, LLR(m)

$$\mathrm{LLR}(m) = \sum_{n' \in N'} [\,\ln \ell(n'|m) - \ln \ell(n'|1)\,]. \qquad (3.3.11)$$

Here, N' are data points within 2 standard deviations from track m.

As we discussed in section 2.1-4, it is possible that the DL process may converge to a wrong solution (which could be a local maximum). In addition to discussion in section 2.4 here we would add the following. Tracks, as other models are pruned and activated as needed. A specifics of tracking problem is that usually a tracking system operates continuously, data are continuously flow in. Therefore if a particular real track is not "captured" after few iterations, it will be captured at a later iteration, after track-model activation. Next, if a spurious track is declared detected, or a real track is missed, these errors will be self-corrected at a later stage of a system operation, when detected track segments or tracklets are connected into longer tracks (this connection of tracklets into longer tracks depends on specific knowledge of a particular system and application requirements; these system operation procedures are outside the scope of this book).

These equations (3.3-6) through (3.3-10) or (2.2-5) through (2.2-7) solve concurrently association and tracking problems. Association is given by f(m|n), and tracking is given by track parameters.

3.3.3 Tracking Example

An application example of the DL tracker is illustrated in Fig. 3.3.1, where concurrent detection (or association) and tracking are performed for targets below the clutter level. Fig. 3.3.1(a) shows true track positions in a 1km * 1km data set, while Fig. 3.3.1(b) shows the actual data available for detection and tracking. In this data, the target returns are buried in the clutter, with signal-to-clutter ratio of about –2dB for amplitude and –3dB for Doppler. Here, the data are displayed such that six radar scans are shown superimposed in the 1km * 1km area, 500 pre-detected signals per scan, and the brightness of each data sample is proportional to its measured Doppler value. Figs. (c)- (h) illustrate the dynamics of the algorithm as it adapts during increasing iterations; the brightness is proportional to association variables, which for this display purpose are computed not just for X(n) but for all pixels (resulting in a smooth image shape). Only association variables for active track models are shown. Fig. 3.3.1(c) shows the initial vague track-model, and Fig. 3.3.1(h) shows track-models upon convergence at 20 iterations. Between (c) and (h) the DL tracker automatically decides how many track-models are needed to fit the data, and simultaneously updates the track parameters and association variables. There are two types of models: one uniform model describing clutter (it is not shown), and linear track-models, which uncertainty changes from large (c) to small (h). In (c) and (d), the DL tracker fits the data with one model, and uncertainty is somewhat reduced. Between (d) and (e) the DL tracker uses more than one track-model and decides that it needs two models to 'understand' the content of the data. Fitting with 2 tracks continues until (f); between (f) and (g) a third track is added. Iterations stop at (h), when similarity stops increasing. Detected tracks closely correspond to the truth (a).

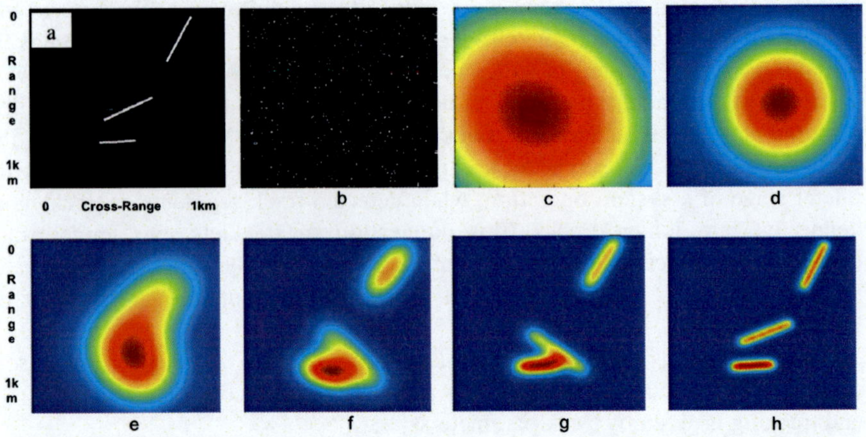

Fig. 3.3.1 Detection and tracking three targets in clutter using DL: (a) true track positions in 1km * 1km data set; (b) actual data available for detection and tracking. DL iterations are illustrated in (c) – (h), where (c) shows the initial, uncertain model and (h) shows the models upon convergence after 20 iterations. Note the close agreement between the converged models (h) and the truth (a).

To summarize, in this example, target signals are below clutter. A single scan does not contain enough information for detection. Detection has to be performed concurrently with tracking, using several radar scans, and six scans are used. In this case, a standard multiple hypothesis tracking, evaluating all tracking association hypothesis, would require about 10^{5000} operations, a number too large for computation. Therefore, existing tracking systems require strong signals, with about a 15db ≈ 30 signal-to-clutter ratio. DL successfully detected and tracked all three targets and required only 10^6 operations, achieving about 18dB ≈ 60 times improvement in signal-to-clutter sensitivity.

3.3.4 Feature Tracking

Feature tracking refers to using features to improve tracking. Features, as discussed in clustering, are some data properties that can be extracted from measurements. For example, a radar cam measure polarization in addition to other data discussed above. Or, a video camera can measure several colors for each pixel. Sometimes features can be used for detecting moving objects, which then are tracked using any tracking algorithm. We concentrate on a more complicated case when features (such as shape) should be extracted from data along with association and tracking. Using standard algorithms this leads to CC for the same reasons as already discussed. Concurrent feature extraction, association, and tracking could be done using DL.

First, we consider a case of features available along with other data. We denote tracking data with features as (**X**(n), **F**(n)); **X**(n) could be GMTI measurements, as in previous sections, or just (x,y) positions, as when a video camera is used. **F**(n) could be any set of available data. In a simplest case, feature tracking combines clustering and tracking. For this simple case, usual uniform clutter similarity can be used, the track models (considering again straight trajectories),

$$\mathbf{M}_m(\mathbf{S}_m,n) = (\mathbf{X0}_m + \mathbf{V}_m T_n, \mathbf{F}_m). \qquad (3.3.12)$$

Here, parameters $\mathbf{S}_m = (\mathbf{X0}_m, \mathbf{V}_m, \mathbf{F}_m, \mathbf{C}_m, r_m)$, where initial positions $\mathbf{X0}_m = (X0_m, Y0_m)$, velocities $\mathbf{V}_m = (V_{mx}, V_{my})$, models of average feature values $\mathbf{F}_m = (F_{m1},\ldots F_{mD})$. Covariances $\mathbf{C}_m$ could be diagonal matrixes $\mathbf{C}_m = \text{diag}(\sigma_{xm}^2, \sigma_{ym}^2, \sigma_{mf1}^2,\ldots \sigma_{mfD}^2,)$, and rates r_m. General DL equations (2.2-5) through (2.2-7) can be used. For images, these equations should be modified as described in section 2.3-1; abs(**X**(n)) in this case can be substituted by abs(**F**(n)). Actually the described here solution is more powerful than standard feature tracking, in that parameters of feature models are estimated concurrently with data association and tracking and with separating moving objects from clutter.

3.4 Swarm Intelligence and Sensor Fusion

3.4.1 Historical Introduction

Thousands of publications and several journals are devoted to these important topics. Reviewing separate approaches or parts of the entire field could be good topics for essays or course reports. Here we just briefly mention several important directions in the field. The simplest case of fusion is when several sensor modalities are co-registered by design. For example, multi-color or hyper-spectral sensors. In this case, fusion just amounts to multi-dimensional clustering, detection, or classification. Much more complicated cases of fusion are when sensors are not collocated. In these cases several sensors might look at overlapping scenes, observing same objects from different angles. If objects of interest can be separated from the rest and identified based on simple measures, such as intensity of signals, colors, location, or velocity (if these measures are available at each sensor) the problem of association of signals with object is easily solved; the problem is then reduced to the previous one.

A particular case of an intermediate complexity is when identification of object location and velocity at every sensor requires tracking objects by each sensor as a first step. Historically, this is a common case of fusion, performed in three steps: detection, tracking, fusion. Even if signal-to-clutter ratio is high, and tracking is relatively easy, still association of signals among sensors may be nontrivial, because of accumulated errors at every step.

Swarm intelligence refers to sensors located on multiple platforms. In these cases, not only sensor information should be exchanged among platforms, but also the behavior of agents (motion, information exchange between agents) is directed at the benefit of the entire swarm. The benefit is measured according to a relevant criterion. Most swarm intelligence algorithms use large swarms and low intelligence. DL lets combining large swarms with highly intelligent agents; in chapter 4 we consider modeling certain aspects of human societies.

We concentrate in this section on the most complex fusion cases, when signals are below clutter, so that concurrent association and tracking is required. We add the next, swarm, level of complexity, when individual sensors do not have sufficient information for solving this problem, and objects can be tracked, identified, and separated from clutter only by using information from several sensors. Also individual platforms know their locations with insufficient accuracy and they should locate themselves relative to each other. More complex issues of swarm behavior, such as navigating multiple platforms and pointing sensors based on shared information for shared goals, we discuss in the Problems section at the end of the chapter. Common algorithms can not solve this type of problems because of CC. And before invention of DL algorithms, problems of this complexity were not considered. DL can solve these type problems with no more difficulty than detection or tracking. When using DL algorithms considered below, one does not have to cut corners by trying techniques designed for simpler cases, first, because clutter is always present and spoils results of simpler algorithms, and second, since DL algorithms, anyway, are often simpler to use and faster to run; and they result in optimal solutions.

3.4.2 Concurrent Localization, Data Association, Navigation, and Fusion for a Swarm of Flying Sensors

Fusion problems can be often solved by extending techniques from tracking and clustering section to multiple sensors. Usually 3-D location and motion of objects in (x, y, z) provide the basis for association, while features from different sensors provide the basis for object identification and separation of objects from clutter. Accordingly, track models should be 3-dimensional, even if sensors measure only two dimensions; e.g. visual and IR sensors measure only 2 angles, radars usually measure range accurately (and sometimes cross-range). However combining measurements from two or more sensors, 3-D location and motion of objects can be reliably estimated. Results are improved if there are phenomenological or physical reasons to develop parsimonious models (with few parameters) predicting all measured features from all sensors, for example color could be the same from all angles. Even so this description is just a short paragraph, the actual description of the models may take many equations. One reason is that when developing models, one has to transform coordinate systems of each sensor into a common coordinate system. These transformations, of course, are simple trigonometric exercise, yet they use long equations. So please keep in mind that these threatening-looking equations are simple and well known, while the real difficulty of associating data and estimating unknown parameters made look simple by using DL.

In this section we consider a problem that combines all sorts of difficulties together, to provide an example of how previously unsolvable problems are made easy and solved using the DL. The problem falls under the broader area of "swarm intelligence." In military surveillance applications, progress in swarm intelligence is expected to revolutionize the ways in which unmanned aerial vehicles (UAVs) are used. The value and potential of UAVs have been demonstrated in recent military conflicts, where they have been used for dangerous and/or tedious missions to reduce the risk of human casualties. It is felt that UAV cooperation and swarming behavior may yield advantages that will make UAVs, in general, even more valuable. The most obvious advantage would be an increase in mission success rates due to improved UAV survivability—hostile defenses would be taxed by the sheer numbers in the swarm. Also, swarms might be deployed in smart ways to increase the efficiency of the geographical coverage. Finally, having access to swarms of sensors may make it easier to detect and discriminate low signal-to-clutter (S/C) targets by exploiting correlations between different, complementary, sensor types and/or different aspect angles.

In order to make deployment of UAV swarms feasible, it will be necessary for UAVs to operate more autonomously than is currently possible. Presently UAVs operate more or less like ''binoculars with wings'' with human operators performing most duties, including low-level functions like image analysis/interpretation and obstacle/collision avoidance. Human operators (and data links from UAVs to operators) would become quickly overwhelmed attempting to control an entire swarm of UAVs. The approach discussed in this section helps reducing the load on human operators by providing computerized interpretation of images from multiple sensors. A by-product of the approach is a set of precise tracks for both targets and UAVs that may be applicable to automatic collision avoidance and for navigation to improve target detection.

One might think it unnecessary to compute UAV positions, since these can be measured directly using onboard inertial devices and global-positioning systems (GPS). However, the accuracy of GPS and inertial measurements may be too rough to allow a particular target's image (signature) in one frame to be reliably associated with its corresponding image in another frame, especially if there are many closely spaced targets, and GPS might be compromised by anti-GPS jamming. For example, the typical accuracy of GPS is on the order of ±10 m [10]. Also, while inertial devices and GPS measure absolute position, they do not measure position relative to potential obstacles or targets. The algorithm described here provides a framework for fine-tuning information from a GPS using outputs from visual (or other) sensors. Thus, in this problem the term "sensor fusion" not only describes combining information from multiple visual sensors, but it also describes combining outputs from visual sensors with outputs from GPS sensors. For optimum performance, all functions need to be performed concurrently: signature association requires accurate UAV tracking, while accurate localization of targets and UAVs requires signature association.

In this example we consider the case in which multiple UAVs, located at the coordinates $\mathbf{X}_j = (X_j, Y_j, Z_j), j = 1,2,\ldots,J$, are flying over a group of objects (''targets'') located at coordinates $\mathbf{x}_k = (x_k, y_k, z_k)$, $k = 1,2,\ldots,K$, where z denotes the elevation and (x,y) denotes the horizontal position (throughout the discussion, vector quantities are indicated in bold type). Note that the term ''target'' is used loosely, referring both to potential threats and simply to landmarks and geographical features to be tracked for the purposes of navigation and registration between multiple images. Each UAV is equipped with an optical sensor (a digital camera which records a matrix of visible or infrared information) and, optionally, a GPS and/or inertial navigation instrument. The GPS measures the UAV position directly, although with significant random error. We denote the coordinate data output by the GPS as $\mathbf{X2}_j = (X2_j, Y2_j, Z2_j)$. Fig. 3.4.1 shows a diagram of one of the UAVs flying over the group of targets.

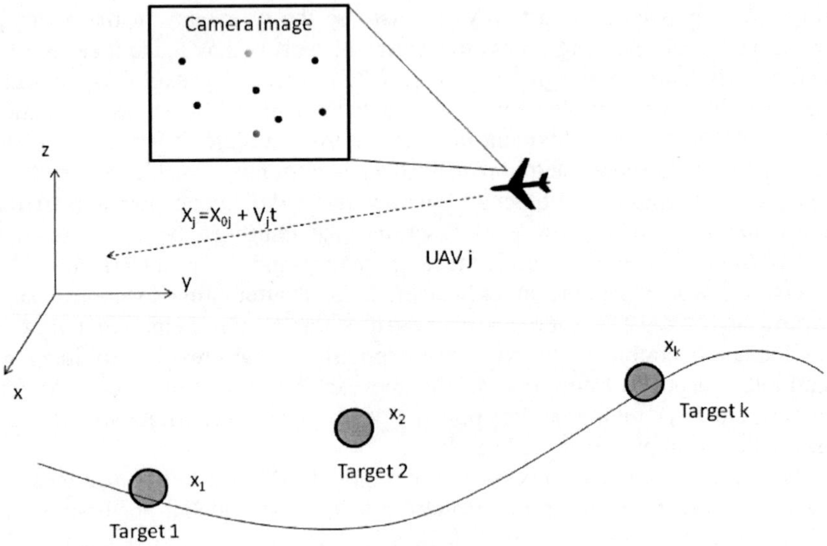

Fig. 3.4.1 a single UAV flying over a group of targets.

The targets are considered to be point reflectors. The sensors on the UAVs record replicas of three dimensional scenes onto two-dimensional images; an object located at $\mathbf{x}_k = (x_k, y_k, z_k)$ is mapped to a (horizontal, vertical) position (a,b) on the camera's focal plane. Because the mapping goes from 3D to 2D, it cannot be reversed to compute a target position uniquely from a single image, even if we know the UAV position. However, from multiple views of the same target it would be possible to triangulate the position, and this illustrates an advantage of having a swarm of sensors. In fact, the problem of localizing objects in 3D based on their image locations in a set of spatially separated photographs is well studied, and is discussed in detail in standard treatments of ''photogrammetry''. Whereas transforming coordinate is a tedious job taking long equations, it does not represent a principal difficulty. The principal difficulty lies in enabling a computer to associate a target signature from one digital photograph with its counterparts in the other photos acquired by other UAVs. This problem is especially acute when the photos contain many targets, some partially obstructed and significant clutter. In the past the problems as difficult as considered here were unsolvable. This ''association problem'' is addressed using DL, as we will discuss below.

The mapping from the 3D world coordinate $\mathbf{x}_k = (x_k, y_k, z_k)$ to the 2D focal plane coordinate (a,b) of a camera located at $\mathbf{X}_j = (X_j, Y_j, Z_j)$ is given by the well known pair of photogrammetric equations

$$a = d_f \frac{(x_k - X_j)m_{11} + (y_k - Y_j)m_{12} + (z_k - Z_j)m_{13}}{(x_k - X_j)m_{31} + (y_k - Y_j)m_{32} + (z_k - Z_j)m_{33}} \qquad (3.4.1)$$

And

$$b = d_f \frac{(x_k - X_j)m_{21} + (y_k - Y_j)m_{22} + (z_k - Z_j)m_{23}}{(x_k - X_j)m_{31} + (y_k - Y_j)m_{32} + (z_k - Z_j)m_{33}} \qquad (3.4.2)$$

where d_f is the camera focal distance, and the quantities m_{rs} are the elements of the 3.3 direction cosine matrix **M** relating the global coordinate frame to the coordinate frame local to the camera on the UAV. Explicitly, the direction cosine elements are given as follows

$$M = \begin{pmatrix} \cos\phi\cos\kappa & \cos\omega\sin\kappa + \sin\omega\sin\phi\cos\kappa & \sin\omega\sin\kappa - \cos\omega\sin\phi\cos\kappa \\ -\cos\phi\sin\kappa & \cos\omega\cos\kappa - \sin\omega\sin\phi\sin\kappa & \sin\omega\cos\kappa + \cos\omega\sin\phi\sin\kappa \\ \sin\phi & -\sin\omega\cos\phi & \cos\omega\cos\phi \end{pmatrix} \qquad (3.4.3)$$

where (ω, ϕ, κ) are the rotational angles (yaw, pitch, and roll) for the coordinate frame of the camera. For simplicity, we will assume these angles can be measured precisely using onboard sensors, although the method can be extended in a straightforward manner to include estimation of (ω, ϕ, κ) along with the other parameters. If we define the vectors as columns in this matrix, $\mathbf{M}_i = (m_{i1}, m_{i2}, m_{i3})$, we can rewrite Eqs. (3.4-1) and (3.4-2) using the compact notation

$$\begin{bmatrix} a \\ b \end{bmatrix} = d_f \begin{bmatrix} M_1 \\ M_2 \end{bmatrix} \frac{(x_k - X_j)^T}{M_3(x_k - X_j)^T} \qquad (3.4.4)$$

where T denotes the vector transpose. As the jth UAV flies, it captures image frames at intervals along its path, and we wish to combine the information from these frames and from the sets of frames from the other UAVs. Models for the UAV flight trajectories will facilitate this task. Consider UAV flying at a constant velocity, UAV j flies with velocity $\mathbf{V}_j$ so that its equation of motion is

$$X_j = X_{0j} + V_j t \qquad (3.4.5)$$

Using this motion model we can rewrite (3.4-4) as

$$\begin{bmatrix} a \\ b \end{bmatrix} = d_f \begin{bmatrix} M_1 \\ M_2 \end{bmatrix} \frac{(x_k - X_{0j} - V_j t)^T}{M_3 (x_k - X_{0j} - V_j t)^T} \tag{3.4.6}$$

The position (a,b) of a target signature in the image is only one piece of the data collected by the cameras. The other piece is the target signature itself, i.e., the array of pixel intensities (red, blue, and green) in the vicinity of the target's image on the focal plane. Most automatic target recognition algorithms make use of a preprocessing step in which a manageable set of classification features are computed from the signature. These features are specially designed to allow signatures from threats to be automatically separated from signatures of clutter objects. In our case, the features will also help in the association problem, as we will discuss. We assume that a set of features $\mathbf{f} = (f_1, f_2, \ldots, f_d)$ has been computed at multiple locations within each image frame.

The data from each target signature include the set of classification features **f** plus the signature location (a,b) on the focal plane. Thus, the information from an image frame is a set of data samples $(a_{jn}, b_{jn}, \mathbf{f}_{jn})$, where n = 1,2,. . . ,N is the index of the sample and j = 1,2,... J denotes which UAV acquired the image. Each of these samples was produced by a particular object (target or clutter). Also recorded with each sample is the time t_{jn} at which the corresponding image frame was acquired. In addition to the data from the camera, we have the data $X2_{jn}$ from the GPS (to make things simple, we assume a GPS data point is acquired simultaneously with each photo). Therefore, the total set of data is contained in the set of samples $w_{jn}=(\mathbf{X2}_{jn}, a_{jn}, b_{jn}, \mathbf{f}_{jn})$ and their corresponding times t_{jn}. Since the rotational angles of each UAV change with time, we will henceforth indicate this dependence in the directional cosine vectors using the notation M_i^{jn} .

At this point we are ready to cast the problem in terms of DL. In previous sections we used shortcut notations n and m for data in pixels n and for models m; here, because of many indexes, we use full notations. Also, we use notations *p* for similarities, ℓ, to emphasize that similarities in this case are pdfs, representing measurements errors (after all parameters are estimated). The data is given as follows

$$w_{jn} = (X2_{jn}, a_{jn}, b_{jn}, f_{jn}), \; j = 1..J \quad n = 1..N \tag{3.4.7}$$

Each data point originates either from some target or from clutter. Thus we need to define two types of models and identify their parameters.

The target model specifies the conditional pdf of the data point w_{jn} coming from target k as follows.

$$p(w_{jn} \mid k) = p_1(X2_{jn}) p_2(a_{jn}, b_{jn} \mid k) p_3(f_{jn} \mid k) \tag{3.4.8}$$

Here the total pdf is broken down into the product of pdf's for the GPS position, the camera coordinates, and the features. This is possible since for each target k these components of the data vector (3.4-7) are independent. We use Gaussian pdf to model sensor errors and thus the three components of the target pdf are expressed as follows.

$$p_1(X2_{jn}) = \frac{1}{(2\pi)^{\frac{3}{2}}\sigma_g^3} e^{-\frac{1}{2\sigma_g^2}(X2_{jn}-MX2_{jn})(X2_{jn}-MX2_{jn})^T} \tag{3.4.9}$$

Where $MX2_{jn}$ is the expected value of the GPS data given by (3.4-5) and σ_g is the GPS error standard deviation. The pdf for camera coordinates is

$$p_2(a_{jn}, b_{jn} \mid k) = \frac{1}{(2\pi)\sigma_a^2} e^{-\frac{1}{2\sigma_a^2}\left[(a_{jn}-Ma_{jnk})^2+(b_{jn}-Mb_{jnk})^2\right]} \tag{3.4.10}$$

where Ma_{ijk} and Mb_{ijk} are the expected values of the camera coordinates computed using (3.4-6), and σ_a is the standard deviation of the error in signature position. Finally, the pdf for the feature data is

$$p_3(f_{jn} \mid k) = \frac{1}{\sqrt{(2\pi)^d \left|C_{fk}\right|}} e^{-\frac{1}{2}(f_{jn}-MF_k)C_{fk}^{-1}(f_{jn}-MF_k)^T} \tag{3.4.11}$$

where C_{fk} is the covariance matrix of the features and d is the number of features. MF_k is the expected value of the feature.

The clutter model is simpler as it describes data points uniformly distributed across the camera focal plane. The model is thus Gaussian over the features and uniform over the other data components. We use the model index k=0 for the clutter model and express the pdf as follows.

$$p\ (w_{jn} \mid 0) = \frac{1}{\sqrt{(2\pi)^d \left|C_{f0}\right|}} e^{-\frac{1}{2}(f_{jn}-MF_0)C_{f0}^{-1}(f_{jn}-MF_0)^T} \tag{3.4.12}$$

All the components of the solution for this problem are summarized Table 3.4-1 below.

The derivatives of the pdf's with respect to all the parameters are obtained using regular calculus. The results of computer simulations are now presented to demonstrate the algorithm.

Throughout these simulations the cameras were assumed to point directly downward. We first considered two examples having four targets distributed within the ground coordinate ranges $[-20 <= (x_k, y_k) <= 20]$, and vertical coordinate $[0 <= z_k <= 10]$, and three UAVs distributed within the ranges $[-30 <= (X_{0j}, Y_{0j}) <= 30]$ and $[15 <= Z_{0j} <= 20]$. The UAV velocities were distributed within the ranges $[-10 <= (dX_j/dt,\ dY_j/dt) <= 10]$ and $[-2 <= d\ Z_j/dt <= 2]$. The full sensor model given by Eq. (3.4-6) was used to calculate the data at time samples t = (0, 1.5, 3) (frame times). For example, in a realistic close-range scenario, all time units might be in seconds and all position units in m.

Table 3.4-1 Components of DL solution for tracking ground targets with multiple UAV's

Component	Description	Notations for Variables/Parameters
Data	GPS position, camera coordinates, image features	$X2_j$, a_{jn},b_{jn}, $\mathbf{f}_{jn}$
Clutter Model	Gaussian feature, Uniform position	C_{f0}, MF_0
Target Model	Gaussian feature, Gaussian GPS, linear motion with Gaussian noise	x_k, X_{oj}, V_j, MF_k, C_{fk}
Known parameters	GPS error and camera alignment error, rates	σ_g, σ_a, r_m

We randomly generated 1600 clutter samples per frame having a single classification feature f with mean MF_0=0 and variance C_{f0}=0.75 . The target features were also randomly drawn from distributions having variance C_{fk} = 0.75 and means of MF_k = [5.5, 7.5, 9.5, 11.5], respectively for k = 1,2,3,4. The K-factor is a commonly used quantitative measure of the degree of separation between two distributions having equal variances. If σ^2 is the variance and ΔM is the separation between the means, then K = ΔM / σ. Thus, for this example the K-factors of each of the four targets vs. the clutter are roughly K = [6, 9, 11, 13]. Also, the standard deviations of the GPS and signature position errors were set to σ_g = 4 and σ_a= 0.1, respectively. Fig. 4a–d shows the results of the simulations, plotted over the space of the UAV 1 sensor focal plane. In (a) the distribution of preprocessed feature data is shown. Here the high values of the target classification features show up as relatively dark pixels over a lighter, speckled, clutter background. For display purposes, the target signatures from all three frame times are shown superimposed onto a single frame of clutter, thus for each of the four targets there are (potentially) three dark pixels corresponding to the three time instances. In (b) the initial, randomly selected, estimates for target signature positions are shown as symbols connected by lines; three symbols for each of four targets. The large circles around signature positions indicate the high initial uncertainty in the estimates.

Plots (c) and (d) show the evolution of the signature position estimates at iterations 10 and 50, respectively. Here, the radii of uncertainty shrink with increasing iterations as the data association becomes less ambiguous. In (d), the data has been properly associated, and the signatures for all four targets have been identified at all frame times.

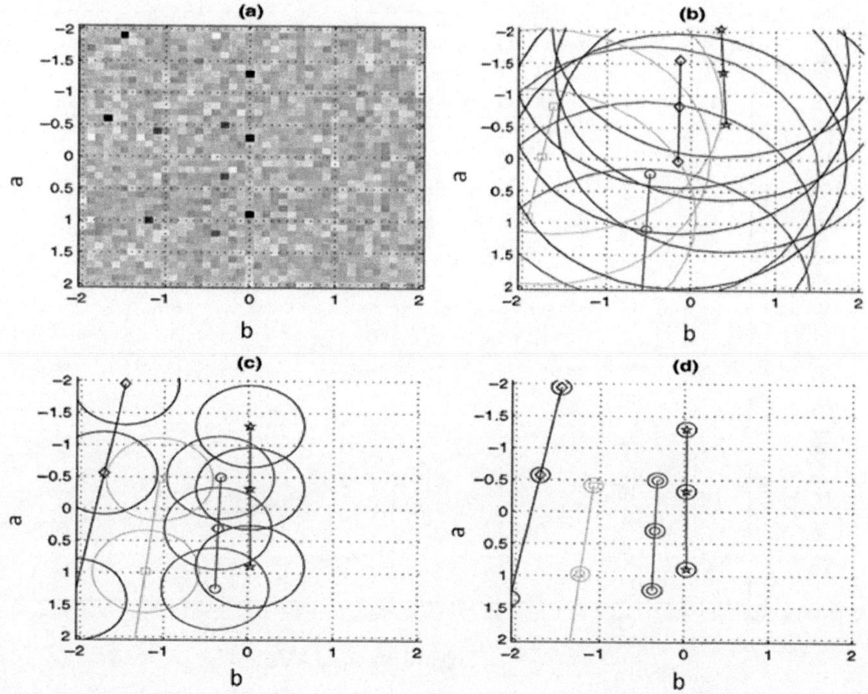

Fig. 3.4.2 Results from the low-clutter example, UAV 1, of 3 total. In (a) the preprocessed feature data is shown distributed over the sensor focal plane. Here the high values of the target features show up as relatively dark pixels over a lighter, speckled, clutter background. In (b) the initial, randomly selected, estimates for target signature positions are shown as symbols connected by lines; three symbols (corresponding to three different time instances) for each of four targets. The large circles around signature positions indicate the high initial uncertainty in the estimates. Plots (c) and (d) show the evolution of the signature position estimates at iterations 10 and 50, respectively. Here, the radii of uncertainty shrink with increasing iterations as the data association becomes less ambiguous.

We generated Monte Carlo results to study the effects of the clutter level on algorithm performance. The error distributions were chosen as in the preceding examples, and target and UAV positions and UAV velocities were generated randomly within the ranges specified above. Figs. 5 plots the errors in estimated target and UAV positions as function of the number of UAVs in the swarm. The vertical axis in these plots indicates the average in radial error, normalized by GPS error, and averaged over 100 Monte Carlo iterations for each data point. From these plots it is apparent that both target and UAV position errors increase roughly linearly with decreasing S/C. Also, the errors decrease as roughly $1/\sqrt{J}$, as J ranges from 2 to 8, where J is the number of UAVs in the swarm.

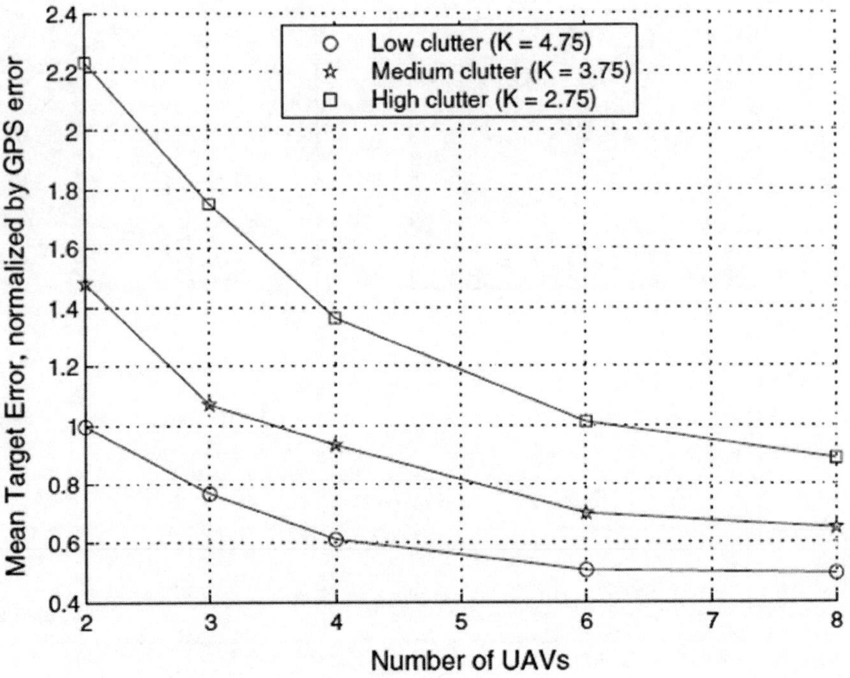

Fig. 3.4.3 Errors in estimated target position vs. signal-to-clutter (proportional to the K-factor) and the number of UAVs in the swarm.

This example illustrates how a complex problem of target detection and tracking can be solved by DL. DL allows to easily combine data coming from different elements of the distributed sensor network, in this case a swarm of UAV's. The solution converges within a limited number of iterations avoiding combinatorial complexity of data association inherent in tracking multiple targets with multiple sensors in clutter.

3.5 Prediction

3.5.1 Linear Regression

A simple approach to prediction is regression. Consider linear regression, when a quantity to be predicted Y is taken as a linear function of known quantities $\mathbf{X}$, = $(X_1, X_2 \ldots X_K)$

$$Y = \mathbf{A}^T\mathbf{X} = A_1X_1 + A_2X_2 \ldots A_KX_K \tag{3.5.1}$$

Coefficients **A** are unknown and should be estimated from examples of Y and **X** known from the past, Y(n) and **X**(n). Solution of this problem is described in hundreds of textbooks and many s/w packages and high-level languages, such as MATLAB, have standard functions to compute regression coefficients **A**. Here we describe the maximum likelihood (ML) approach to solving of this problem, for comparison with solutions of more complex problems that could be obtained with DL. Below we shorten the argumentations and present the gist of the ML derivation, while leaving the full ML derivation for Problems.

The ML approach to linear regression assumes that Y(n) and **X**(n) are random realizations of (Y,**X**), which have Gaussian pdf of the following shape,

$$\mathrm{pdf}(Y,\mathbf{X}) = (1/2\pi\sigma^2)^{1/2} \bullet \exp[-(Y - \mathbf{A}^T\mathbf{X})^2/2\sigma^2]. \quad (3.5.2)$$

This expression for the pdf corresponds to our standard notation $\ell(n)$. One interpretation of this distribution is that the difference $(Y - \mathbf{A}^T\mathbf{X})$ is due to measurement errors, and errors often follow Gaussian distributions. Correspondingly, one maximizes the likelihood function L over parameters **A**,

$$L = \prod_{n \in N} (1/2\pi\sigma^2)^{1/2} \bullet \exp[-(Y(n) - \mathbf{A}^T\mathbf{X}(n))^2/2\sigma^2]. \quad (3.5.3)$$

Instead of maximizing L, it is easier to consider lnL, which is equivalent to minimizing a sum of square:

$$\min \sum_{n=1...N} (\,Y(n) - \mathbf{A}^T\mathbf{X}(n)\,)^2. \quad (3.5.4)$$

For this reason using a fundamental statistical principle of the ML is equivalent in this case to "sum of square minimization," or Least Mean Square.

To make equations shorter, we change variables. We introduce notations <…> for averages (similar to eq. (3.3-5))

$$<\ldots> = (1/N) \sum_{n=1...N} <\ldots>. \quad (3.5.5)$$

Instead of (Y(n), **X**(n)), we consider

$$(Y'(n), \mathbf{X'}(n)) = (Y(n), \mathbf{X}(n)) - <(Y(n), \mathbf{X}(n)>. \quad (3.5.6)$$

This results in the following equations:

$$0 = \sum_{n=1...N} [(Y'(n) - \mathbf{A}^T\mathbf{X'}(n))]\mathbf{X'}(n)^T. \quad (3.5.7)$$

We denote

$$\mathbf{B}^T = < Y'(n)\mathbf{X'}(n)^T >, \quad (3.5.8)$$

and an estimated covariance

$$\mathbf{C} = < \mathbf{X'}(n)\mathbf{X'}(n)^T>; \quad (3.5.9)$$

we obtain a solution of (3.5-7) for regression coefficients $\mathbf{A}^T$,

$$\mathbf{A}^T = \mathbf{B}^T\mathbf{C}^{-1}. \quad (3.5.10)$$

Of course, every software package contains procedures for matrix inversion (computing $\mathbf{C}^{-1}$), equivalently for solving a linear systems of equations (3.5-7). The estimation in this simple case of a single Gaussian distribution (a single regression process) does not require an iterative DL procedure. Combining this with eqs.(3.5-1) and (3.5-6), the regression equation is obtained as

$$Y = \langle Y \rangle + \mathbf{A}^T(\mathbf{X} - \langle \mathbf{X} \rangle). \tag{3.5.11}$$

3.5.2 Example of Linear Regression

Fig. 3.5.2. illustrates simple linear regression in 2-D. The LMS fit is shown with a line and the 2D Gaussian fit is shown with the ellipse. It is clear that both methods give the same result.

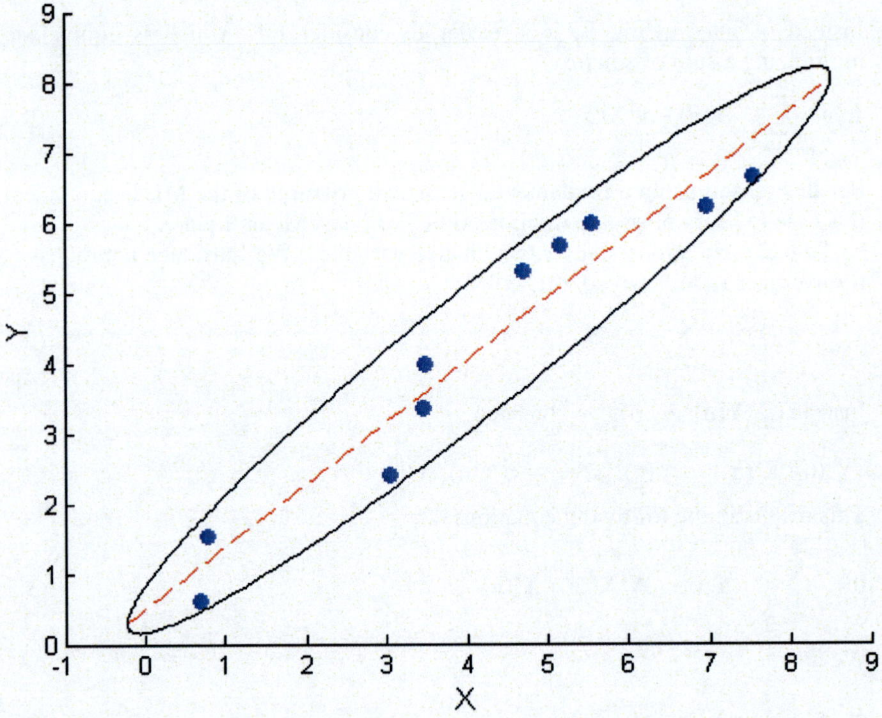

Fig. 3.5.2 Linear regression. The data points (X Y) are produced by Gaussian density with mean value of (3.5 3.5) and covariance matrix (3 0.9; 0.9 1). The dashed line shows the linear fit to the data. The ellipse illustrates the Gaussian density.

3.5.3 DL Regression in Clutter

The previous example of regression in presence of clutter, was selected for an illustration purpose so that the clutter point was obvious. In the past such an obviously wrong data point would thrown out as an outlier. Various ad-hoc rules would be used for more complex cases, say consider as an outlier every data point that is more then 3 standard deviations away, etc. This approach is not quite satisfactory, because it requires first estimating wrong regression and standard deviation from data containing clutter. It will not work if there is more clutter points than “good” data points; how to decide what clutter is. Here we consider a model for linear regression + clutter and develop a DL estimation, which separates clutter points from real data points in the “best” way.

We consider a standard uniform conditional similarity for clutter model, m=1,

$$\ell(n|1) = 1 / \text{volume}(Y,\mathbf{X}) = 1 / [\, (\max Y - \min Y) \prod_d (X_{max}-X_{min})_d \,]. \qquad (3.5.13)$$

Note, in (3.5-2) and (3.5-3) we considered only one variable, $(Y - \mathbf{A}^T\mathbf{X})$. Here, we account for clutter that can have any value in $(Y,\mathbf{X})$ space. Correspondingly, we need similarities defined in this entire space. We consider first an extension of eq.(3.5-3) to a uniform distribution of **X**, similar to (3.5-13): A regression conditional similarity for model m=2 differs from (3.5-3) by accounting for **X**,

$$\ell(n|2) = (1/2\pi\sigma_2^{\ 2})^{1/2} \bullet \exp[-(\,Y(n) - a_2 - \mathbf{A}_2^{\ T}\mathbf{X}(n))^2/2\sigma_2^{\ 2}] \,/ \prod_d (X_{max}-X_{min})_d\,. \qquad (3.5.14)$$

Another difference here is coefficient a_2; it accounts for the fact that we do not know average values of $(Y,\mathbf{X})$ under each hypothesis; by maximizing similarity we can only estimate a_2, which is the average value of $Y - \mathbf{A}_2^{\ T}\mathbf{X}$, under the m=2 hypothesis.

In statistics, a standard linear regression eq.(3.5-1) with the Gaussian pdf (3.5-2) is considered as the mean (average) value of Y, given **X**. Similarly, statistical description of a liner regression in clutter is given by a mixture model pdf (the two components of the mixture describe the regression process and the clutter process),

$$\text{pdf}(Y,\mathbf{X}) = r_1 / \text{volume}(Y,\mathbf{X}) + r_2\, (1/2\pi\sigma_2^{\ 2})^{1/2} \bullet \exp[-(Y - a_2 - \mathbf{A}_2^{\ T}\mathbf{X})^2/2\sigma^2] \,/\, \text{volume}(\mathbf{X})\,. \qquad (3.5.15)$$

Here, volume(**X**) is a product in (3.5-14). Conditional pdfs correspond to (3.5-13) and (3.5-14). For this pdf the mean value of Y, given **X**, is

$$Y = r_1\, \text{pdf}(\mathbf{X}\,|1) / \text{pdf}(\mathbf{X}) + r_2\, \text{pdf}(\mathbf{X}\,|2) / \text{pdf}(\mathbf{X}) \bullet \mathbf{A}_2^{\ T}\mathbf{X}. \qquad (3.5.16)$$

In terms of the standard DL notations this corresponds to

$$Y(n) = f(1|n) + f(2|n) \bullet [a_2 + \mathbf{A}_2^{\ T}\mathbf{X}(n)]. \qquad (3.5.17)$$

This equation can be interpreted as follows. If **X**(n) fits well into the regression model prediction,

$$Y(n) = a_2 + \mathbf{A}_2^T\mathbf{X}(n), \qquad (3.5.18)$$

then f(2|n) >> f(1|n), and Y is predicted according to the standard regression model (3.5-18); otherwise, **X**(n) is interpreted as clutter, and Y(n) is predicted accordingly. Thus, handling clutter occurs automatically.

Parameters (a_2, $\mathbf{A}_2$, r_1, r_2, σ_2) are estimated using a standard DL procedure, eqs.(2.2-5 – 2.2-7). An alternative estimation procedure can start, as usually, with any parameter values (σ_2 should be defined large enough), and the following iterations are repeated until convergence:

$$(1)\ f(m|n) = r_m \ell(n|m) / \sum_{m' \in M} r_{m'} \ell(n|m'), \qquad (3.5.19)$$

$$(2)\ r_m^{it+1} = (1/N) \sum_{n \in N} f(m|n), \qquad (3.5.20)$$

$$(3)\ a_2 = < Y(n) - \mathbf{A}_2^T\mathbf{X}(n) >_2; \text{ where } < (\ldots) >_2 = (1/N) \sum_{n \in N} f(2|n)\ (\ldots)_n \qquad (3.5\text{-}21)$$

$$(4)\ \mathbf{B}_2^T = < (Y(n) - a_2)\ \mathbf{X}(n)^T >_2;\ \ \mathbf{C}_2 = < \mathbf{X}(n)\mathbf{X}(n)^T >_2; \qquad (3.5.22)$$

$$(5)\ \mathbf{A}_2^T = \mathbf{B}_2^T\mathbf{C}_2^{-1}, \qquad (3.5\text{-}23)$$

$$(6)\ \sigma_2^2 = < Y(n) - a_2 - \mathbf{A}_2^T\mathbf{X}(n) >_2; \qquad (3.5.24)$$

In these equations $< (\ldots) >_2$ is the same as in eq.(3.3-5). So, on each iteration, the regression estimation equations in steps (3, 4, 6), eq. (3.5-21), (3.5-22), and (3.5-24), differ from (3.5-8, 3.5-9) by substituting weighted sums instead of the plain sums over n. Therefore, regression coefficients, $\mathbf{A}_2^T$, are estimated predominantly from those data points that fit the regression equation (3.5-18), if the regression predicts Y much better than clutter does.

When value of Y is absent and the regression eq.(3.5-17) is used for predicting Y, f(m|n) are evaluated using the expected values of Y under the corresponding hypothesis. For m=1, f(1|n) does not depend on Y; for m=2, f(2|n) is evaluated using Y predicted according to the regression (without clutter) eq. (3.5-18). In this case exp in (3.5-15) equals 0, and (3.5-17) can be made more specific. Denote

$$\Delta_1 = (\max Y - \min Y); \quad \Delta_2 = (2\pi\sigma_2^2)^{1/2}. \qquad (3.5.25)$$

Then, we obtain

$$f(1|n) = (r_1/\Delta_1) / [(r_1/\Delta_1) + r_2/\Delta_2] = \Delta_2 r_1 / [\Delta_2 r_1 + \Delta_1 r_2];$$
$$f(2|n) = (r_2/\Delta_2) / [(r_1/\Delta_1) + r_2/\Delta_2] = \Delta Y r_2 / [\Delta_2 r_1 + \Delta_1 r_2];$$

So that the predicted value of Y according to (3.5-17) is

$$Y = [\Delta_2 r_1 + \Delta_1 r_2 (a_2 + \mathbf{A}_2^T\mathbf{X}(n))] / [\Delta_2 r_1 + \Delta_1 r_2]; \qquad (3.5.26)$$

If clutter and regression error are low, $r_1 << r_2$, and $\Delta_2 << \Delta_1$, the predicted value of Y is close to the standard linear regression.. Note, that even in a strong clutter, if a regression error is low, $\Delta_2 r_1 << \Delta_1 r_2$, regression coefficients are accurately estimated.

Instead of uniform similarities of (Y,**X**) for the clutter model, and **X** for the regression model, one can use Gaussian, or other models. This will only result in slightly different values of r_1 and r_2. But it might require estimating more parameters, which increases estimation errors; therefore it is only justified if there are serious reasons (an a priori knowledge of probability densities).

3.5.4 Example of DL Regressions in Clutter

This example demonstrates the effect of outliers on linear regression. The data in Fig. 3.5.3A is the same as in Fig. 3.5.2 with the exception of a single data point (0 4). Even though this point is an obvious outlier the LMS approach includes it in the estimation resulting in incorrect linear fit. The DL approach captures the outlier by the clutter model and results in correct estimate as illustrated in Fig. 3.5.3.B.

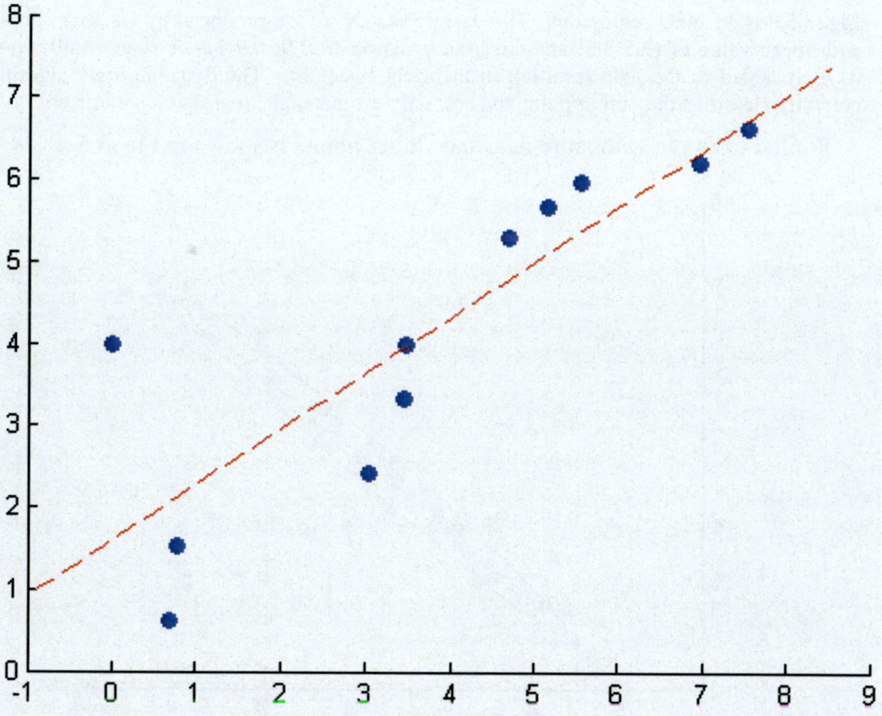

Fig. 3.5.3 (A) Linear regression. The data points (X Y) are produced by Gaussian density with mean value of (3.5 3.5) and covariance matrix (3 0.9; 0.9 1). A single outlier point (0 4) is added to the data resulting in incorrect linear fit.

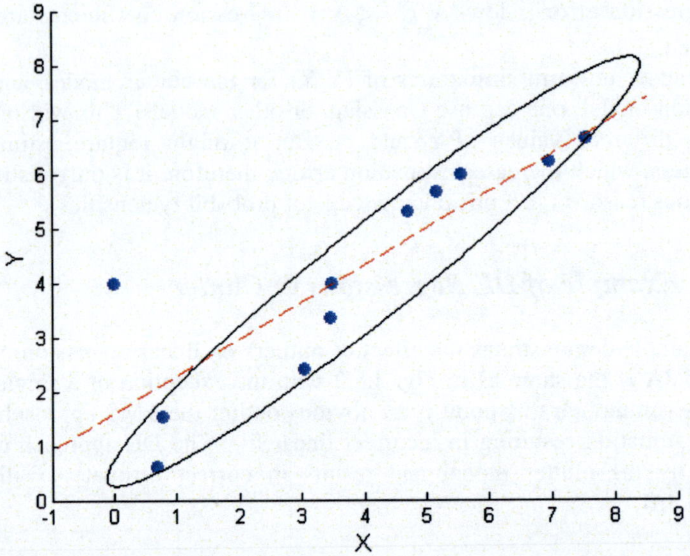

Fig. 3.5.3 (B) Linear regression. The data points (X Y) are produced by Gaussian density with mean value of (3.5 3.5) and covariance matrix (3 0.9; 0.9 1). A single outlier point (0 4) is added to the data resulting in incorrect linear fit. The dynamic logic algorithm correctly classifies the outlier point and correctly estimates the regression parameters.

Similar example with more data and clutter points is shown in Fig. 3.5.4.

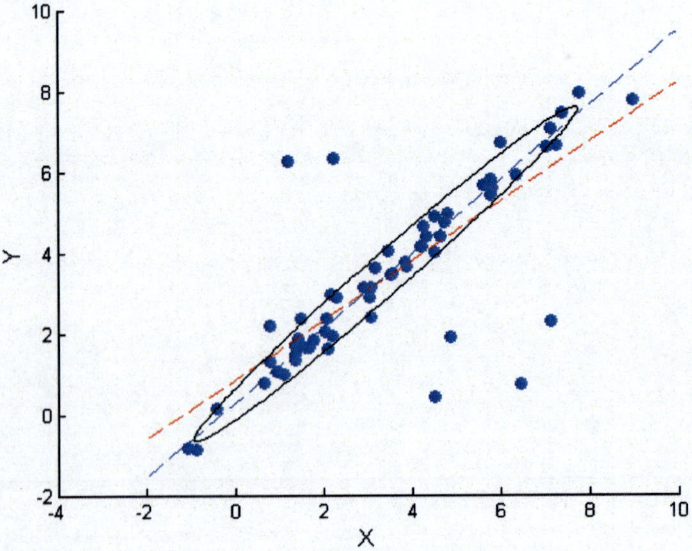

Fig. 3.5.4 Linear regression with noisy data. The data points (X Y) are produced by a mixture of Gaussian density with mean value of (3.5 3.5) and covariance matrix (3 0.9; 0.9 1) and a normal density. The dynamic logic algorithm correctly classifies the data point and correctly estimates the regression parameters. The linear fit is shown in red and the correct linear fit is shown in blue.

3.5.5 Multiple DL Regressions

In addition to clutter, several regression processes can be active at once. A particular example we consider later in financial predictions. The procedure in the previous section is readily modified. Instead of (3.5-17), the multiple regression equation is a weighted sum of several linear regressions

$$Y(n) = \sum_{m \in M} r_m \, f(m|n) \bullet [\, a_m + \mathbf{A}_m{}^T \mathbf{X}(n) \,]. \qquad (3.5.27)$$

As usual, the first model is clutter, so that $a_1 = 1$, $\mathbf{A}_1 = \mathbf{0}$. Estimation follows either the standard DL procedure or eqs.(3.5-19) through (3.5-24) extended to multiple regression models $m > 2$.

This might be useful, e.g. when index n parallels time, and a system switches from one control or regression law to another.

3.6 Financial Prediction

This section discusses some basic knowledge necessary for quantitative financial predictions. At least, you will learn how to avoid widely occurring errors, and how not to lose money needlessly. There are few (or better to say, very very few) quantitative trading systems that consistently outperform broad financial markets. Of course everyone understands, that an owner of such a system will not publish it in all details for everyone to use. And you should not expect to find it here. Still dynamic logic is so much more powerful than other "state of the art" techniques, that we can share with readers potentially useful approaches, which we do not have time to explore ourselves. It would take a lot of effort to turn these ideas into profitable quantitative trading systems. We are sure that if any of our readers will seriously embark on this path, he or she will invite us as consultants; this is of course true for any application described in this book.

The objective of financial prediction is to optimize portfolio, which is a balance between maximizing growth and minimizing risk. We discuss various appropriate measures of growth and risk. We discuss approaches to developing quantitative algorithms for predicting markets, and procedures for training and evaluating these algorithms.

Financial predictions are different from other types of predictions, such as weather, or most physical phenomena. The difference is that millions of people in the world are trying to make money by predicting directions of financial markets. Therefore prediction techniques that are well known are used by many traders, including large financial institutions, which execute trades very fast, ahead of individual traders. Everything that is predictable in financial markets *by using well known techniques* has been already traded out and cannot be used by individual investors for making gains. Because of this financial markets are considered "efficient." Many financial textbooks, and even exams for highly prestigious designations (such as CFA, Chartered Financial Analyst) state an 'efficient financial market hypothesis': it is impossible to make better than average gains by trading in financial markets.

Of course this is not true, which is proven by existence of people such as P. Lynch, G. Soros, W. Buffett, and others, who consistently over decades produced better than average gains by trading in financial markets. Therefore, financial textbooks (and exams) formulated a "soft version of efficient financial market hypothesis," which acknowledges that better than average gains are only possible by having better than average analytical skills. Still those aspiring for CFA are not allowed to answer the corresponding exam question by saying that better than average gains are possible by having better than average mathematical prediction techniques. At least this was the case when one of this book authors took CFA exams.

Please, take a note, this section is written for mathematicians, not for professional traders, who trade other people's money. *Trading other people's money is a highly regulated business and is prohibited by law in the USA (and in many other countries) without having proper licenses.* Legal professional requirements are not discussed in this book. As a matter of disclosure: we do not trade other people's money, otherwise we would be legally prohibited from saying some of the things we say below.

3.6.1 Testing Procedure

There are good reasons for the denying that "better than average mathematical prediction techniques" exist. First, such techniques are indeed difficult to come by. Second, there are many people, who use usual mathematical techniques, known by many; still they convince themselves that they can produce better than average gains. Of course, when they start actual trading they quickly lose money. Therefore, we start not with financial prediction algorithms, rather this section concentrates on what is an appropriate testing procedure for any financial prediction technique.

The first step is to decide which financial instrument one would like to trade, and which data to use in prediction algorithms. Data on many financial instruments are available on the Internet for free. The second is to develop a prediction technique; assume it was done. This section discusses how to test this prediction technique.

Definition of portfolio rate of return. Denote p(t) a portfolio value at time t. We assume no money are added or taking out of portfolio, and the only changes in value of the portfolio are due to trading securities, receiving dividends and interest, paying trading commissions and fees, and changing in security values. The rate of return of a portfolio from time t_1 to t_2 is

$$R = (p(t_2) - p(t_1)) / p(t_1). \tag{3.6.1}$$

Usually it is expressed in %. The rate of return in portfolio should be compared to a standard market benchmark portfolio, e.g. S&P500. It is usually computed over equal time intervals, dt (which could be a second, a day, or a week, depending on how often trades are made. A rate of return computed from (t-dt) to t, we denote

$$dR(t) = (p(t) - p(t-dt)) / p(t-dt). \quad (3.6.2)$$

Total or cumulative rate of return (3.6-1) can be expressed as a product of incremental rates of return (3.6-2):

$$R = (p(t_2) - p(t_1)) / p(t_1) = [1+dR(t_1+dt)] \bullet [1+dR(t_1+2dt)] \bullet \ldots \bullet [1+dR(t_2)] - 1. \quad (3.6.3)$$

Because financial markets change their behavior over time, prediction algorithms should be adaptive. That is they are "trained" using data from some past period. For example, parameters of a regression algorithm are estimated using past data, say over the last year to predict the next day, or the next week. It is good to include periods of growing markets as well as periods of declining markets into the training data. Before starting actual trading, one would like to know how good the algorithm is. So one can test the algorithm, say over the last 3 months, and compare the cumulative gain to that of the benchmark portfolio.

This simple procedure, however, is inadequate. Since, the last 3 months were included into the training interval (the last year), the algorithm was *fitted* to the training data. It is surprising how sensitive is testing procedure to this fact. A testing interval should be *strictly* excluded from the training interval. If I plot my simulated cumulative gain (or portfolio value) over the test interval, and I see a wonderful curve smoothly rising, the first thing I'll suspect is that somehow training and testing intervals overlapped.

Let us say, you verified carefully, that training and testing intervals never overlapped. If you tested from t_2+dt to t_3; your training was from t_1+dt to t_2.as you moved your testing interval forward to the current day (say by 1 day at a time), you also moved the training interval by one day at a time and they never overlapped. You looked at your results and they are not as good as you expected. You decide that you need to change few things (a length of training interval; or to add one more predictive variable, say interest rates in Australia; and delete interest rates in Hong Kong). Say, results became better. You tweaked a bit more and results are real good. Beware – this procedure is similar to fitting your algorithm to your data, similar to overlapping training and testing data.

Therefore, algorithms should be first developed and tweaked from t_1+dt to t_2 (say, t_2 was two years ago). Then you test it from t_2+dt to t_3 (say, t_3 was one year ago). Possibly you made one or two last tweaks. Then, finally you test it from t_3+dt to t_4 (and this is today) if it works all right, possibly you are ready to trade. Still, trade for a month or more on paper, account for commissions and fees and for slippage (say you ordered 'buy' on your computer at $X, but it might be actually exercised at $X + 0.05%. Now compare your results to the benchmark.

3.6.2 Three-Process Model for Financial Prediction

One way to develop algorithms for financial predictions is to model types of financial traders who dominate the markets. Here we model two widespread types of traders: people who trade on momentum, and people who trade on value. We consider daily close in a particular market (say S&P 500) as X, and denote a daily change dX(t) = X(t)-X(t-1). On majority of days the market recent behavior may not be consistent for trading decisions, and the 1st model, m=1, we model as clutter (no trades). Momentum traders (m=2) try to catch on a developing trend. If the market predominantly moves in a particular direction, this could be indicated by an average of dX over several days, say D_2 days, or just by the difference X(t) – X(t-D_2). It might be a good idea to take more recent information with a stronger weight than older data. Therefore we model a D_2-day trend indicator by using so called exponential moving average over D_2 days:

$$EMD_2(t) = \sum_{d=1...D} \exp(-(d-1)/D_2)\, dX(t-d+1) \,/ \sum_{d=1...D} \exp(-(d-1)/D_2). \qquad (3.6.4)$$

If $EMD_2(t) > th_2$ the momentum trader buys, if $EMD_2(t) < -th_2$, the momentum trader sells:

Momentum model: if $EMD_2(t) > th_2$, **buy**; if $EMD_2(t) < -th_2$, **sell**. (3.6.5)

The threshold value, th_2, could also be made a model parameter to be learned from data; buy or sell thresholds do not have to be symmetrical..

A value trader, model m=3, tries to sell an overvalued market and buy an undervalued one. Warren Buffett did it for decades by reading financial statements of the companies, by talking to their management teams, and sometimes by changing the management team after buying the company. For our purpose we use a purely "technical" indicator that is, computed from recent market valuations (prices, X(t)). If the market price in a short run significantly outperforms the market long-term performance, the value trader considers the market overvalued, and the other way around. A short and long term market behavior is modeled using $EMSD_3(t)$ and $EMLD_3(t)$; these are computed using eq.(3.6-4) with parameters SD_3 and LD_3 in place of D_2.

Value model: if $EMLD_3(t) > EMSD_3(t) + th_3$, **buy**; if $EMLD_3(t) < EMSD_3(t) - th_3$, **sell**. (3.6.6)

Again, the threshold value, th_3, could be a model parameter to be learned from data; buy or sell thresholds do not have to be symmetrical. One can start with, e.g $LD_3 = 2 \cdot SD_3$.

Now we discuss estimation of the parameters of a trading system based on these models, using our standard DL for maximizing similarity. These parameters are:

$$r_1, r_2, r_3, \sigma_1, \sigma_2, \sigma_3, D_2, SD_3, LD_3, th_2, th_3. \qquad (3.6.7)$$

For simplicity and shortness, we consider zero thresholds, th_2, th_3 = 0. First we select N days, which we use for training our models (instead of our traditional index n, here we use t). Verify that the training interval encompasses several up and down markets, e.g. $(X_{max}-X_{min}) > 3 \bullet abs(X(N)-X(1))$. Then, let us formulate conditional similarities corresponding to the 3 models discussed above (clutter, momentum, value); the trade rules above we consider as predictors or models, M_m, of the next-day price change, dX; e.g., under model m=2 we consider $M_m(t) = EMD_2(t-1)$ as a predictor of dX(t). The three models are given by

$$M_1(t) = 0; \quad M_2(t) = EMD_2(t\text{-}1); M_3(t) = EMLD_3(t\text{-}1) - EMSD_3(t\text{-}1). \qquad (3.6.8)$$

Correspondingly, conditional similarities are defined as follows

$$\ell(t|1) = (X_{max}-X_{min}), \qquad (3.6.9)$$

as usually, we do not consider X_{max} and X_{min} as model parameters; we compute them directly from the data; For models m = 2,3,

$$\ell(t|m) = (1/2\pi\sigma_m^2)^{1/2} \bullet \exp[-(dX(t) - M_m(t))^2/2\sigma_m^2] . \qquad (3.6.10)$$

Using these equations and the standard DL estimation procedure, parameters (3.6-7) are estimated.

Using these parameters, the dX(t) prediction, Y(t), is estimated similarly to eq. (3.5-27)

$$Y(t) = \sum_{m \in M} r_m f(m|n) \bullet M_m(t). \qquad (3.6.11)$$

This gives the three-model trading rule,

Three-model rule: if Y(t) > 0, **buy**; if Y(t) < 0, **sell**. (3.6.12)

It is a good idea to make a simple order-of-magnitude test. Verify that the sum of all Y(t) is on the order of the sum of all dX(t), which is the same as the total market change over the training interval, $\Sigma Y(t) \sim (X(N) - X(1))$. If these quantities significantly differ, verify your computer code for errors (see section 2.4).

The above model (3.6-12) does not optimize for the percentage of portfolio used for each trade. If one uses all available cash on a 'buy' signal, the next trade has to wait until a 'sell' signal, and v.v. It could result in too few trades, or too many trades, depending on your personal temperament. Portfolio management is not just a matter of mathematical optimization. You have to be comfortable with your model rules, so that after the model development is finished, you can actually faithfully follow recommendations. Let us discuss few approaches to tailoring the trade rule to your personal preferences. One can use a probabilistic rule, trading proportionately to the strength of the Y(t) signal. A simple rule can be developed as follows. Compute a standard deviation of Y(t) over the training interval,

$$\sigma_Y = [(1/N) \sum_{t \in N} Y(t)^2]^{1/2}. \qquad (3.6.13)$$

On every 'buy' or 'sell' signal, trade the percentage of cash or securities available in the portfolio, 'buy' using $b \bullet (Y(t)/\sigma_Y) \bullet$ cash; or 'sell' $s \bullet (Y(t)/\sigma_Y) \bullet$ securities. (3.6.14)

Parameters b and s can be selected using your own preferences by few trials over the training interval and then optimized around the selected values. Similarly, one can exercise trades only if recommendations (3.6-14) exceed a threshold, th. Again, th can be selected according to your preferences, and then optimized near the selected value. How to optimize portfolio is discussed in the next section.

3.6.3 Portfolio Optimization

Optimization of portfolio is not simply maximizing gain over the training period. In addition, usually, one would also reduce risk, and possibly keep the number of trades within certain range. All of these criteria could be combined in a single value function, which is optimized during training. A simple and important measure of risk is portfolio standard deviation, STD, computed using the rate of return (3.6-1, 3.6-2):

$$STD = [(1/N) \sum_{t \in N} (dR(t) - dR_{average})^2]^{1/2}; \quad dR_{average} = R/N. \tag{3.6.15}$$

Using this measure of risk, one could optimize a ratio of return to risk. A more important measure, called Sharpe ratio, is a ratio of return to risk computed using portfolio return relative to a risk-free asset return, R_f (such as return on a 3-month T-bill). Substituting $(R-R_f)$ and $(dR-dR_f)$ in the above equations, Sharpe ratio is given by

$$S = (R-R_f) / STD_f. \tag{3.6.16}$$

Here STD_f is a standard deviation of the portfolio rate of return vs. risk-free rate of return. For S&P 500 typical Sharpe ratio is about 0.4.

To maximize Sharpe ratio, while keeping the number of trades near the desired value, $N_{desired}$ one can penalize Sharpe ratio for the deviation of the number of trades N_{tr} from its desired value, as follows

$$V = S - a\,(N_{tr} - N_{desired})^2, \tag{3.6.17}$$

and maximize this function V during training.

3.7 Situational Awareness, Context Understanding

3.7.1 DL for Learning Situations

Learning situations is a next step beyond pattern recognition, in complexity of the problem and solution. In pattern recognition, tracking, and other applications considered above, the DL processes from vague to crisp begin with uncertain model parameters and vague associations of models with data. For learning situations in this section, vagueness has to be understood on a different level.

Since situations are collections of objects, sets, vagueness of sets has to be defined to apply DL. For intelligent agents acting in real-world identifying objects and identifying situations occurs in parallel, or almost in parallel; here we ignore this system-level aspect and consider learning situation as a separate step undertaken after objects have been recognized.

The principled difficulty for learning situations from objects is that every situation includes many objects that are not essential to recognition of this specific situation; in fact there are many more "irrelevant" or "clutter" objects than relevant ones. Let us dwell on this for a bit. Objects are spatially-limited material things observed by sensors. A situation is a collection of contextually related objects that tend to appear together and are perceived as meaningful, e.g., an office, a dining room. The requirement for contextual relations and meanings makes the problem mathematically difficult. Learning contexts comes along with learning situations; it reminds the problem of a chicken and egg. The human mind subliminally perceive many objects, most of which are irrelevant, e.g. a tiny scratch on a wall, which we learn to ignore. To formulate this process mathematically, using DL, vague models of sets have to be defined. As we describe in this section, the DL extension to situations and contexts turned out to be straightforward.

The total number of objects that a system can recognize in the world we denote Do (it is a large number). In every situation an agent perceives D_p objects. This is a much smaller number compared to Do. A situation could be a clutter situation, containing D_p random objects, or a meaningful situation characterized by the presence of D_s objects essential for this situation ($D_s < D_p$).

We number situations by n, and represent a situation mathematically as a vector in the space of all objects, $X_n = (x_{n1}, \dots x_{ni}, \dots x_{nDo})$. If the value of x_{ni} is 1, the object i is present in the situation n and if x_{ni} is 0, the corresponding object is not present. Since D_o is a large number, X_n is a large binary vector with most of its elements equal to zero. Situation models, as usually are numbered by m; each model is characterized by parameters, a vector of probabilities, $p_m = (p_{m1}, \dots p_{mi}, \dots p_{mDo})$. Here p_{mi} is the probability of object i being a part of the situation m. Thus a situation model contains D_o unknown parameters. Estimating these parameters constitutes learning situations.

The parameters, elements of vector $\mathbf{p}_m$ we model as independent (this is not essential for learning, if presence of various objects in a situation actually is correlated, this would simplify learning, e.g. perfect correlation would make it trivial). Correspondingly, conditional similarity of observing vector $\mathbf{X}_n$ in a situation *m* is then given by the standard expression called binomial distribution,

$$\ell(n|m) = \prod_{i=1}^{Do} p_{mi}{}^{X_{ni}}(1 - p_{mi})^{(1-X_{ni})}. \quad (3.7.1)$$

This conditional similarity is vague if all p_{mi} are close to 0.5, and every object belongs (or does not belong) to every situation with approximately 50% probability. The similarity is crisp, when all p_{mi} are close to 0 or 1; so that every object definitely belongs or does not belong to a situation with 100% or 0% probability.

Among N situations observed by the agent most observations are "irrelevant," corresponding to observing random sets of objects (clutter), and there are M-1 "real" situations, in which D_s objects were repeatedly present (M and D_s are not known, D_o and D_p are known, since they are actually observed and recognized). All random observations we model by 1 model (clutter); the probabilities for this clutter model, m=1, are taken p_{1i}=0.5 for all *i*. Thus we define M possible sources for each of the N observed situations.

For shortness, we do not discuss relations among objects. Spatial, temporal, or structural connections, such as "to the left," "later," or "connected" can be easily added to the above DL formalism. Relations also require markers (indicating which objects are related). Relations and corresponding markers re no different mathematically than objects, and can be considered as included in the above formulation. An alternative computational mechanism accounting for relations is the hierarchy; relations could be models at a higher level, combining related objects in a higher-level structure. Both type of relations, "flat" and hierarchical, have to appear in a self-learning system as a result of learning. We address learning of hierarchies in the next chapter.

The formulation here assumes that all the objects have already been recognized, but the above formulation can be applied without any change to real, continuously working agent with multiplicity of concurrently running DL processes at levels of objects and situations, feeding each other. The object signals can be sent to DL-situation processes before objects are fully recognized, while DL-object processes are still running and object representations are vague; this would be represented by x_{ni} values between 0 and 1. The presented formalization therefore is a general mechanism for agents learning objects and situations.

The general DL equations (2.2-5, 2.2-6, 2.2-7) can be used to estimate the p_{mi} parameters (M and D_s come out automatically, as demonstrated in the next section). However, iterative equations for p_{mi} can be simplified:

$$p_{mi} = [\sum_{n\in N} f(m|n)\, x_{ni}] \,/\, [\sum_{n'\in N} f(m|n')] \,. \qquad (3.7.2)$$

3.7.2 Example of Situation Learning

In this example we set the total number of recognizable objects equal to 1000 (D_o=1000). The total number of objects perceived in a situation is set to 50 (D_p=50). The number of essential objects is set to 10 (D_s=10). The number of situations to learn (M-1) is set to 10. Note that the true identities of the objects are not important in this simulation so we simply use object indexes varying from 1 to 1000. The situation names are also not important and we use situation indexes. The data for this example are generated by first randomly selecting D_s=10 specific objects for each of the 10 groups of objects, allowing some overlap between the groups (in terms of specific objects). This selection is accomplished by setting the corresponding probabilities p_{mi} = 1. (Note that eq. (3.7-1) have numerical problems for p_{mi} = 1 or 0, therefore we actually vary them between 0.05 and 0.95). Next we add 40 more randomly selected objects to each group (corresponding to D_p=50). We also generate 10 more random groups of 50 objects to model situations without specific objects (noise); this is of course equivalent to 1 group of 500 random objects. We generate N'=800 observations for each situation resulting in N=16,000 situation observations (data samples, n = 1... 16,000) each represented by 1,000-dimensional vector $\mathbf{X}_n$. These data are shown in Fig. 3.7-1 sorted by situations.

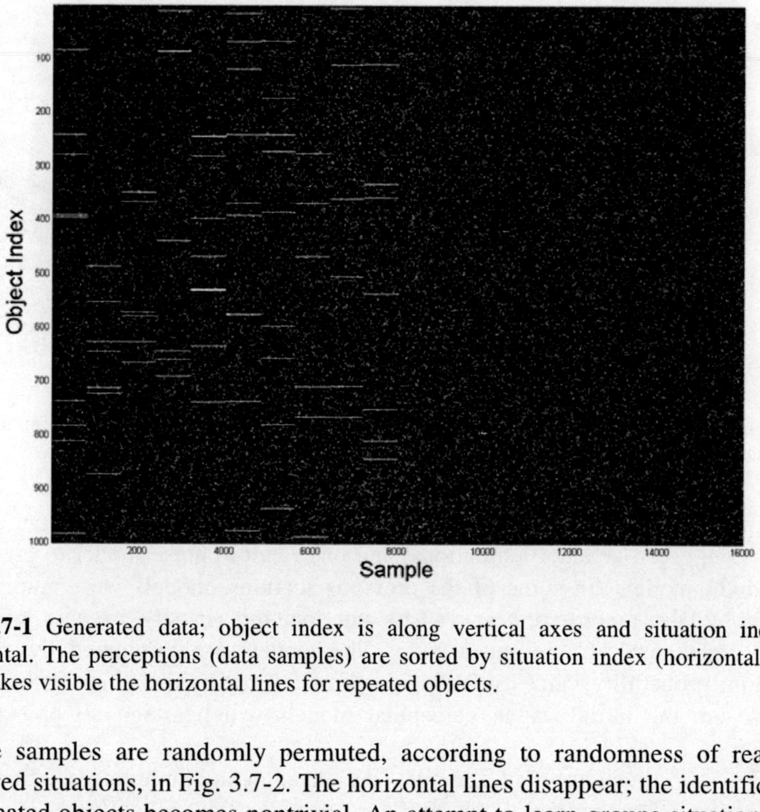

Fig. 3.7-1 Generated data; object index is along vertical axes and situation index is horizontal. The perceptions (data samples) are sorted by situation index (horizontal axis); this makes visible the horizontal lines for repeated objects.

The samples are randomly permuted, according to randomness of real life observed situations, in Fig. 3.7-2. The horizontal lines disappear; the identification of repeated objects becomes nontrivial. An attempt to learn groups-situations (the horizontal lines) by inspecting various horizontal sortings (until horizontal lines would become detectable) would require $M^N = 10^{16000}$ inspections, which is of course impossible.

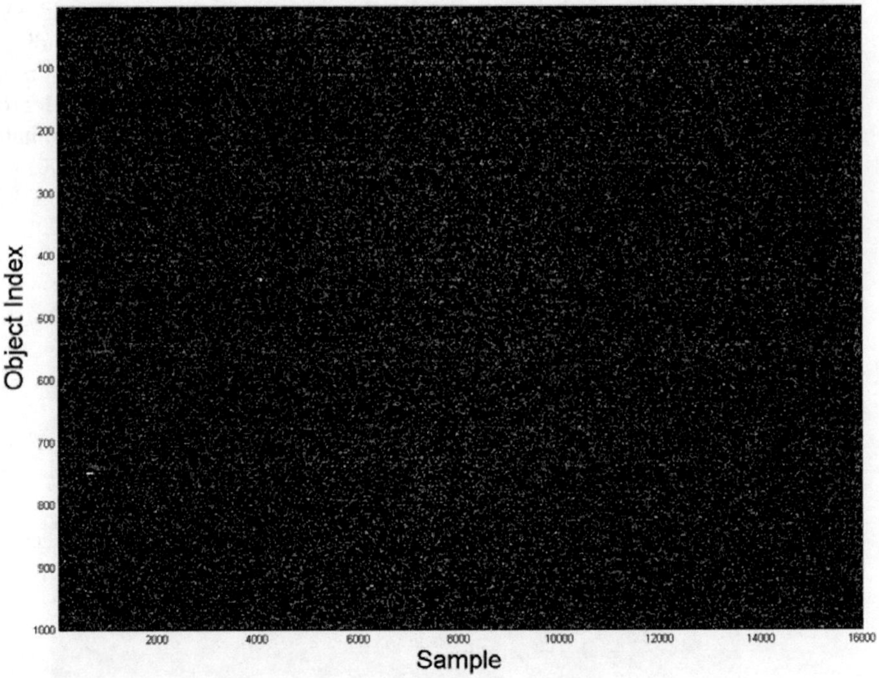

Fig. 3.7-2 Generated data, same as Fig. 3.7-1, randomly sorted by situation index (horizontal axis), as available to the DL algorithm for learning.

The DL algorithm is initiated by defining 20 situational models (an arbitrary selection, given actual 10 situations) and one random noise model to give a total of M=21 models (in some of the previous sections, models were automatically added by DL as required; here we have not done this (mostly, because it would be too cumbersome to present results). The models are initialized by assigning random probability values in the vicinity of 0.5 to the elements of the models. These are the initial vague perceptual models which assign all objects to all situations.

Fig. 3.7-3 illustrates the DL models initialization and iterations (the first 3 steps of solving DL equations. Each subfigure displays the probability vector $\mathbf{p_m}$ for each of the 20 models. The vectors have 1000 elements corresponding to objects (vertical axes). The values of each vector element are shown in gray scale. The initial models assign nearly uniformly distributed probabilities to all objects. The horizontal axes are the model index changing from 1 to 20. The clutter model is not shown. As the DL algorithm progresses, situation learning improves, and only the elements corresponding to repeating objects in "real" situations keep their high values, the other elements take low values. By the third iteration the 10 situations are identified by their corresponding models. The other 10 models converge to more or less random low-probability vectors (clutter).

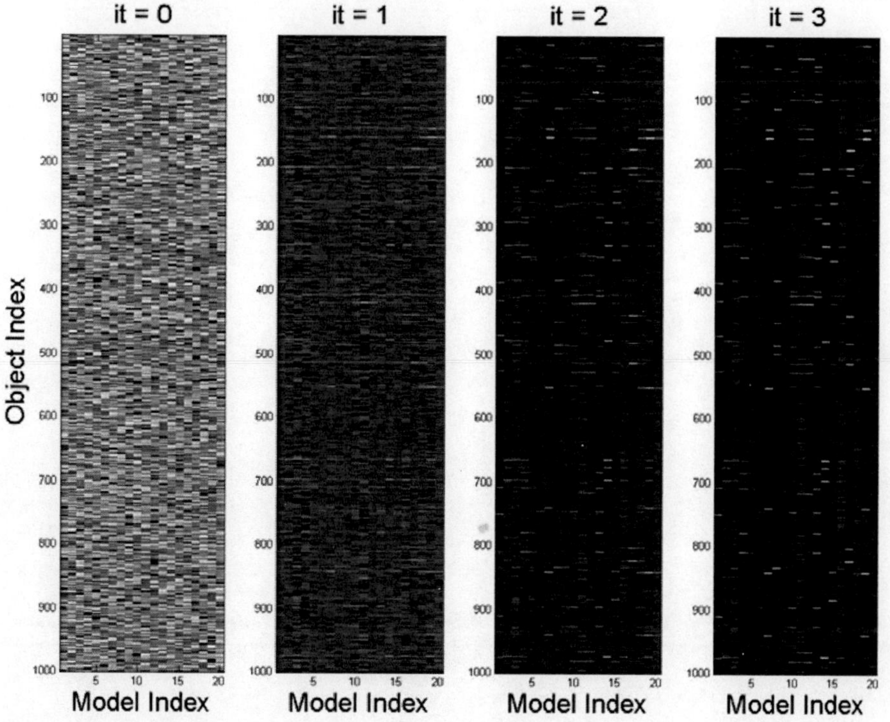

Fig. 3.7-3 DL situation learning. Situation-model parameters converge close to true values in 3 steps.

This fast and accurate convergence can be clearly seen in Figs. 3.7-4 and 3.7-5. We measure the fitness of the models to the data by computing the sum squared error, using the following equation.

$$E = \sum_{m\in\{B\}} \sum_{i=1}^{Do} (p_{mi} - p_{mi}^{True})^2 .$$

In this equation the first summation is over the subset {B} containing top 10 models that provide the lowest error (and correspondingly, the best fit to the 10 true models). In a real-time agent, of course, the best models would be added as needed, and the random samples would accumulate in the noise model automatically; as mentioned, DL can model this process and the reason we did not model it, is that it would be too cumbersome to present results. Fig. 3.7-4 shows how the sum squared error changes over the iterations of the DL process. It takes only a few iterations for the DL to converge. Each of the best models contains 10 large and 990 low probabilities. Iterations stop, when average error of probabilities reached a low value of 0.05 resulting in the final error E(10) =1000*(0.05^2)*10 = 25.

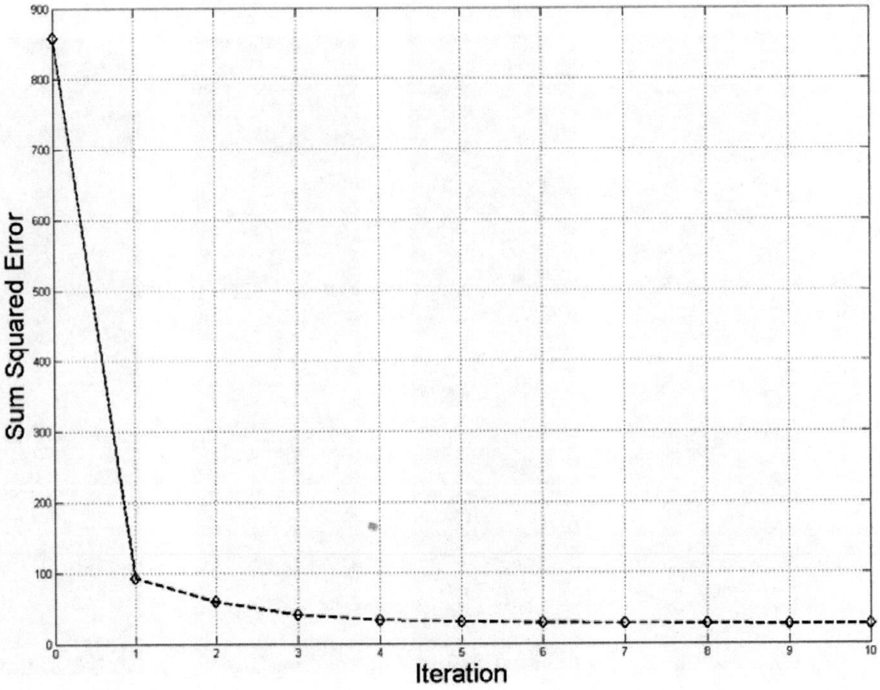

Fig. 3.7-4 Errors of DL learning are quickly reduced in 3-4 steps, iterations continue until average error reached low value of 0.05 (10 steps).

Fig. 3.7-5 shows average associations, A(m,m') among true (m) and computed models (m'); this is an 11x11 matrix according to the true number of different models (it is computed using association variables between models and data, f(m|n))

$$A(m,m') = (1/N') \sum_{n=1}^{N} f(m|n)^* \, f(m'|n), \; m' \in \{B\}, \tag{3.7.3}$$

$$A(m,11) = (1/10^*N') \sum_{m' \notin \{B\}} \sum_{n=1}^{N} f(m|n)^* \, f(m'|n), \; m' \notin \{B\}. \tag{3.7.4}$$

Here, f(m|n) for true 10 models m is either 1 (for N' data samples from this model) or 0 (for others), f(m'|n) are computed associations, in the second line all 10 computed noise models are averaged together, corresponding to one true (random) noise model. The correct associations on the main diagonal in Fig. 3.7-5 are 1 (except clutter model, which is spread among 10 computed clutter models, and therefore equals 0.1) and off-diagonal elements are near 0 (incorrect associations, corresponding to small errors shown in Fig 3.7-4.)

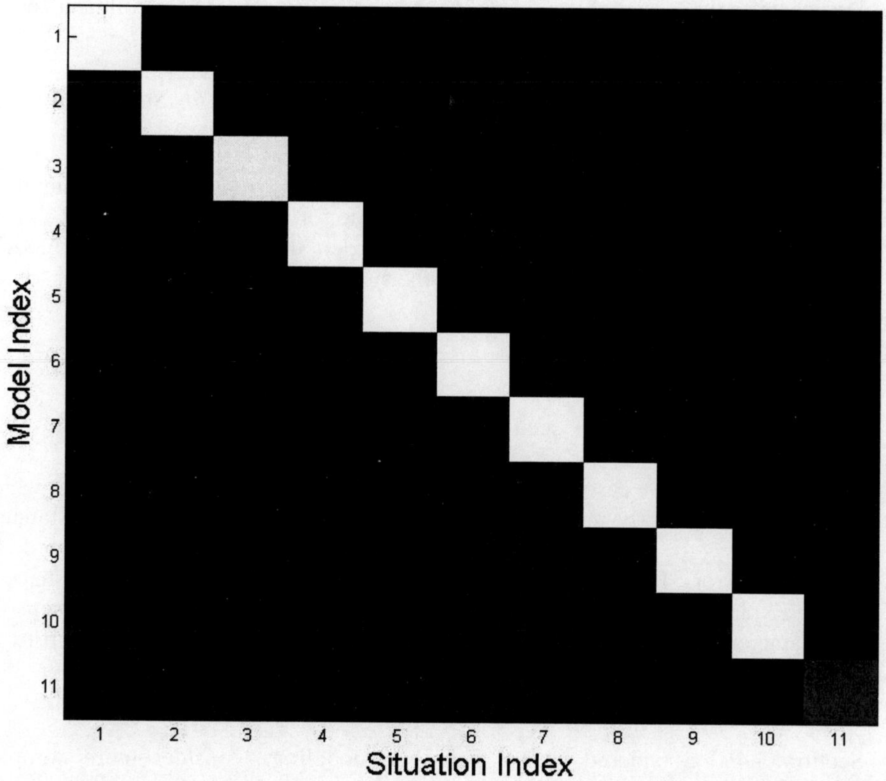

Fig. 3.7-5 Correct associations are near 1 (diagonal, except noise) and incorrect associations are near 0 (off-diagonal).

3.8 Problems (*Master Thesis Level, +PhD Thesis Level)

P3.8.1 Detect Circular Shapes in Clutter*

Follow example in section 3.1. Simulate few circular shapes using Gaussian functions. Add random clutter. Detect circles as in section 3.1. Repeat for 1 circle, while changing two parameters (1) strength of clutter (signal-to-clutter ration SCR = r(circle)/ r_1); and (2) distance, d, between the true circle center and the initial model circle center (make sure that the initial radius of the model is larger than d). Publish results.

P3.8.2 Regression Equation Problems (for Students Inclined toward Theoretical Derivations).

Prove that the mean values of Y and **X** under each hypothesis (m) cannot be estimated by maximizing similarity. Hint: define for each m $(Y', \mathbf{X'})_m = (Y, \mathbf{X}) - (\underline{Y}, \underline{\mathbf{X}})_m$; here $(\underline{Y}, \underline{\mathbf{X}})_m$ are the means under the mth hypothesis (model).

Demonstrate that equations maximizing the similarity (3.5-15) are collinear for a_m, and for any component of $(\underline{Y}, \underline{X}^d)_m$;

P3.8.3 Repeat 3.8.1 for Regression Equation Problems (for Students Inclined toward Theoretical Derivations).*

Prove that the mean values of Y and **X** under each hypothesis (m) cannot be estimated by maximizing similarity. Hint: define for each m $(Y', \mathbf{X}')_m = (Y, \mathbf{X}) - (\underline{Y}, \underline{\mathbf{X}})_m$; here $(\underline{Y}, \underline{\mathbf{X}})_m$ are the means under the mth hypothesis (model). Demonstrate that equations maximizing the similarity (3.5-15) are collinear for a_m, and for any component of $(\underline{Y}, \underline{X}^d)_m$;

P3.8.3 Tracking†*

*Implement algorithms in section 3.3.4 and compare to performance of other algorithms.

†A more complex case: use shape-related and other features for targets, which values are not directly measured by a sensor. A simple case could use standard deviations (σ_{xm}, σ_{ym}) as features; if the target is unresolved, these values would be just sensor errors. For resolved targets, they would give target extents and shape in (x, y). Explore various possibilities of this approach for various types of sensors, for characterizing two and three dimensional target shapes, and other properties.

P3.8.4 Swarms and Fusion†*

Section 3.4.1 considered swarm behavior modeling. Consider more complex issues of swarm behavior, such as navigating multiple platforms and pointing sensors based on shared information for shared goals.

*P3.8.5 Compare Bayesian and Information Similarities *†*

Compare solutions obtained by using 2.2 (Bayesian) and 2.3 (information) similarities – for clustering, for tracking, and for other applications. PhD work would require theoretical derivations, e.g. computing Cramer-Rao Bounds (CRB, see Perlovsky 1989, it would require some changes) for simple cases and comparing to simulations similar to P3.8.1. Publish several papers.

P3.8.6 Recognition of Objects as Situations†

Modify algorithm for learning situations from section 3.7 for learning objects. Model objects as situations of features, limited in its space size (space size limits can be implemented by using Gaussian distributions in space.

3.9 Literature for Further Reading

3.9.1 Clustering

[Perlovsky, 2001] contains detailed analysis of using DL for clustering with Gaussian Mixture Models (GMM) and comparison to other technique, in particular, to K-Nearest Neighbors (KNN) [Fukunaga, 1972]. [Perlovsky, 2001] discusses extensions of DL from unsupervised (clustering) to supervised (classification), to partially supervised, and to supervised with probabilistic teacher applications.

Early publications on clustering with Gaussian Mixture Models (GMM) by using DL include [Perlovsky 1987, Perlovsky & McManus 1991]. Before these publications, GMM were not widely used; they where considered too complex, and their convergence was thought as problematic. A pioneer in pattern recognition and cluster analysis, K. Fukunaga, in the first edition of his book [Fukunaga, 1972] suggested KNN method for clustering as the best one. After extensive discussions with one of the authors in 1987-1988, in which DL with GMM was demonstrated to outperform KNN significantly, Dr. Fukunaga included GMM into the second edition of his book [Fukunaga, 1990]. This discussion is still of the current interest, as many authors are using KNN for clustering. Recent discussion of clustering with GMM can be found in [Xu & Wunsch, 2008]. Also see Duda, Hart, and Stork 2000.

3.9.2 Tracking

Kolmogorov published a solution to tracking problem in 1941. Norbert Wiener solved tracking problem, apparently independently from Kolmogorov during the 1940s and published in (Wiener, 1949). The technique is called Wiener-Kolmogorov filter. It treated the problem of measurement noise, but did not consider clutter (extraneous signals). It is appropriate for a target on linear trajectory, in absence of any other signals.

Kalman filter (1960) could use complex models. However, similar to Wiener-Kolmogorov filter it did not consider association problem and therefore could only track a single object without clutter. A number of algorithms were developed, which attempted to add association to Kalman filter, Nahi 1969; Jaffer & Bar-Shalom 1972. A general MHT tracking algorithm Singer, Sea & Housewright, 1974; Reid, 1979; Blackman, 1986; Probabilistic Data Association tracking algorithms, PDA and JPDA and related algorithms Bar-Shalom & Tse 1975; Fortmann, Bar-Shalom & Scheffe 1980; Streit & Luginbuhl 1994; Willett, Ruan, & Streit 2002; Ruan & Willett 2004.

3.9.3 Swarm Intelligence and Sensor Fusion

Deming & Perlovsky 2007, Hall & Llinas 2001.

3.9.4 Situations and Contexts

Endsley 1995; Ilin & Perlovsky 2009; Perlovsky & Ilin 2009.

Chapter 4
Emerging Areas

The brain-mind is a most sophisticated and advanced system, solving many problems, which cannot be solved today by computers. Therefore modeling the brain-mind mechanisms is a broad emerging area of engineering, which goals are to develop cognitive algorithms approaching human intelligence. We begin this chapter with discussing experimental evidence for dynamic logic modeling actual perception and cognition processes in the brain. This helps understanding the brain-mind mechanisms and provides a new foundation for understanding successes of the DL algorithms described earlier. Based on this relation between DL and brain mechanisms, as well as recent progress in cognitive neuro-psychology, more complicated brain processes are modeled including grounded symbols and language learning. Correspondingly more complicated emerging areas of engineering are described including natural language learning, search engines based on language understanding, the role of emotions in cognition, building cooperative man-machine systems, in which man and machines learn from each other and form smoothly cooperating systems. This requires modeling higher cognitive abilities, including the beautiful, music, sublime. Possibly a most challenging problem of the 21st century is to understand cultural diversity, to understand differences and similarities among diverse cultures. Cognitive, mathematical, and engineering understanding of the diversity will help us to accept it, to learn how to live together in the diverse world.

Complex cognitive, cultural, and social mathematical models discussed here will be developed using the mathematics of interacting intelligent agents; each agent mathematical model will follow algorithms in chapter 3, especially section 3.7; several new fundamental mathematical ideas will be introduced throughout this chapter. Let us emphasize that also some sections in this chapter have few equations, nevertheless, even most esoteric sections concerning the beautiful and music give detailed outline for combining mathematical techniques from other sections into societies of intelligent agents mathematically modeling human societies, cultures, their emotions, and their evolution.

Throughout this book we relegated literature discussion to a separate section at the end of each chapter. This chapter is different; we often discuss references in the text. The reason is that this chapter discusses emerging areas where literature discussion and ongoing research are essential. We address areas on the boarder between linguistics, cognitive science, psychology, mathematical modeling of the brain, and engineering. These areas, let us repeat, stir much controversy and address many unsolved problems. It is appropriate therefore to address the central

L. Perlovsky et al.: Emotional Cognitive Neural Algorithms with Eng. Appl., SCI 371, pp. 81–174.
springerlink.com

issues of these controversies, the remaining unsolved problems, and literature discussions, alongside with discussing how the proposed theory offers solutions to these complex problems and outlines future research directions.

This chapter reviews emerging areas of mathematical modeling and engineering by unifying mathematics with psychology and cognitive science. Much of the most advanced research in physics during the 20th century, from electromagnetism to quantum superstrings has been moved by a vision of "grand unification," a theory that would unify all elementary forces of nature. We believe that the most fruitful and interesting science of the 21st century will address mathematical modeling of the mind. Through this "grand unification" of mathematics and sciences of the mind much will be gained for deeper understanding of the mind-brain as well as for engineering of intelligent systems. Every section and subsection in this chapter guides to novel areas in cognitive science, and to mathematical approaches to developing algorithms and engineering systems. Every one of these areas is a fruitful field for future research, for many papers, books, Master and Ph.D. Theses.

4.1 Fundamental Mind Mechanisms

Instincts, concepts, emotions, bottom-up and top-down signals, hierarchy, consciousness.

Instincts are among most ancient mechanisms of the mind. Historically, in psychological literature, basic instinctual mechanisms were mixed up with "instinctual behavior." This intuitive idea combines a lot of complicated mechanisms, which could not have been scientifically understood or mathematically modeled. Facing this complexity, psychologists abandoned the idea of instincts, and today this word is not popular among psychologists, instead psychologists talk about drives, motivations, and other less clear notions, which are not scientifically defined or modeled.

We return to an intuitively clear idea of instinct, which we define clearly, succinctly, in correspondence with physiological data, and in a way that can be easily modeled mathematically. An instinct is a sensor-like mechanism that measures a basic organism need. We have dozens of such sensors in our body, measuring chemistry, fluid and bodily pressures around our bodies; most are acting autonomously and unconsciously. Some of the most important for survival produce more or less clear, consciously perceived signals. For example, we have sensors measuring a sugar level in blood, when sugar is low, we feel hunger, we want to eat. In this book we are not interested in physiology, but in mathematical modeling, therefore defining instincts as sensors is sufficient for our purpose. Moreover, in this book we are not interested in details of functioning of the body. We are interested in functioning of the mind, in acquiring knowledge; therefore we study in details the knowledge instinct (KI). This is a sensor-like mechanism that measures knowledge, a correspondence of our ideas to reality, and drives us to improve the knowledge. Below we discuss that KI mathematical model is similarity described in chapter 2; we discuss that KI is a most important instinct driving our higher mental abilities.

The mind understands the world due to the mechanism of concepts or mental representations. Mental representations model sensor signals from objects, situations, and events in the world. Therefore mathematical models of mental representations, in a simplified way, are just mathematical models of objects and events in the world, similar to models discussed in chapters 2 and 3, and often they are called simply mental models. Later we discuss models appropriate to represent various mechanisms of the mind.

4.2 Dynamic Logic and Cognition

Mathematical modeling of concepts, imagination, intuitions, instincts, emotions, including aesthetic emotions, and emotion of the beautiful.

4.2.1 Dynamic Logic, Concepts, Hierarchy, and Unconscious

The brain-mind is organized as a multi-level approximately hierarchical system. We start with considering mechanisms of visual perception. Because we concentrate on the general principles rather than specific neural mechanisms, we will refer to *mind* rather than *brain*. At lower levels the mind perceives primitive elements of the visual situation. At higher levels these elements are organized into objects. Next, objects are organized into situations. Still higher in the hierarchy are abstract cognitions. Primitive elements, objects, situations, abstract cognitions are called mental representations and are modeled mathematically by models similar to those we considered in previous chapters. This hierarchical structure is illustrated in Fig.4.1-1.

This architecture modeling brain-mind using DL was also called historically neural modeling fields (NMF), but in this book we mostly keep a uniform designation, DL. At every level, mental representations-models are created by interacting neural signals coming from in opposite directions. Neural signals coming "up" the hierarchy are called bottom-up signals, and signals coming "down" are called top-down signals. Bottom-up signals are generated by mental representations created or excited at lower levels, and top-down signals are generated by higher-level representations. The hierarchy is approximate, there are essential interactions among non-adjacent levels, still we will use "hierarchy" for simplicity. At the very bottom of the hierarchy, bottom-up signals are generated by sensor organs. We talk about visual system for the definiteness; so the sensor organs are eyes, and neural signals are generated by eye retinas.

Consider first mathematical description of the mind at the level of objects. Mental representations-models of objects can be considered similar to models in section 3.1. Later we discuss where these models have "come from." Now we discuss that the process of dynamic logic, considered in previous chapters, models a fundamental process of perception *"from vague-to-crisp."* Look at an object in front of you, a book, or computer, or pen. Then close eyes and imagine this object. The imagined object is not as crisp and clear as perception of this object with opened eyes. During the recent 15 years neuroscientists learned that the imagined object is created by top-down projections (signals) from mental representations (models) of this object (Grossberg, Kosslyn). The fact that the imagined object is vague attests to vagueness of mental representations, similar to initial states of models in the DL process in Fig. 3.1.1.c. Recently similar experiment was repeated with much more details using neuroimaging techniques (Bar et al 2006). It was found that perception of a familiar object takes about 150 to 180 ms. During this time the brain matches the initial vague top-down image to the crisp bottom-up image projected to the visual cortex. (Actually, the bottom-up image is also changed in this process, but we would not concentrate on this now.) This process is unconscious, it takes place outside of consciousness. We are only conscious about the final crisp perception. These crisp conscious states of the mind correspond to what is normally called perceptions, cognitions, or *concepts*.

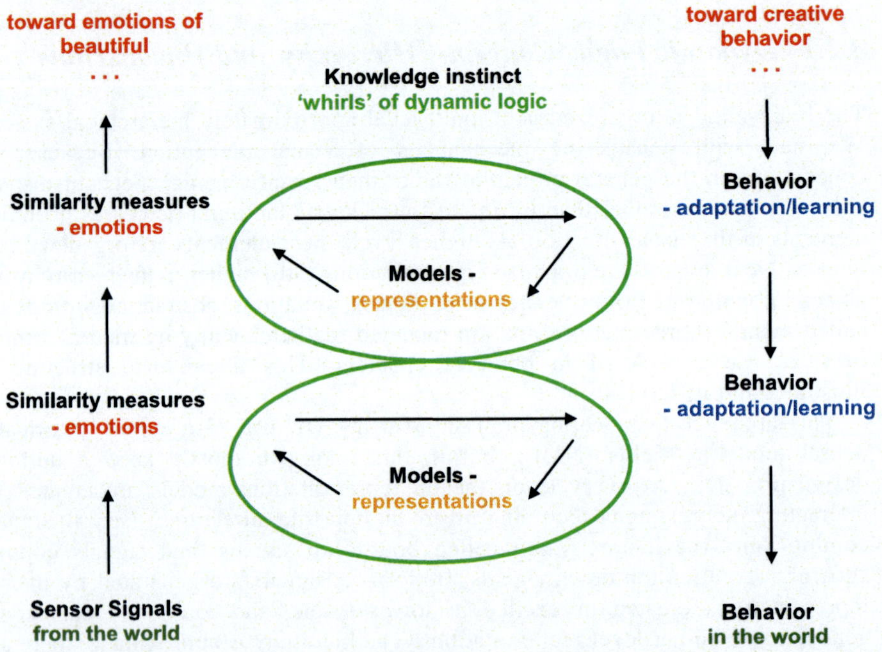

Fig. 4.1-1 Hierarchical NMF system. Sometimes term "heterarchy" refers to cross-level connections, not shown, and to the consequence that the hierarchical structure is not logically strict as may appear from the figure. At each level of the hierarchy there are models, similarity measures, and actions (including adaptation, maximizing the knowledge instinct - similarity). Concept-model activations are output signals at this level and they become input signals to the next level, propagating knowledge up the hierarchy. The hierarchy is not strict; interactions may involve several levels. At the top of the hierarchy there are models of meaning and purpose, related emotions of beautiful, and creative behavior.

A "concept" designates a common thread among words such as concept, idea, understanding, thought, or notion. Concepts are abstract in that they treat individual entities as if they were identical. Emphasizing this property, medieval philosophers used the term "universals," Plato and Aristotle called them ideas or forms, and considered them the basis for the mind understanding of the world. Similarly, Kant considered them a foundation for the ability for understanding, the contents of pure reason [40]. According to Jung, conscious concepts of the mind are learned on the basis of inborn unconscious psychic structures or archetypes [39]. Contemporary science, as mentioned, equates the mechanism of concepts with internal representations of objects, their relationships, situations, etc. In NMF concepts are described by models, $\mathbf{M}_m$. The essential mechanism of DL, as discussed, is the process "from vague to crisp," models stored in memories are vague, fuzzy, uncertain; during perception and cognition they generate initial top-down signals; in interactions with bottom-up signals models become concrete, certain, and crisp.

We are not conscious about firings of individual neurons in our brain. Similarly, the entire perception process "from vague-to-crisp" is not accessible to our consciousness. Nevertheless, our consciousness "convinces" us that we perceive objects immediately, as soon as we look at them. Tens to hundreds thousands of neurons participate in the perception of a simple object. Only may be 0.01% of this neural activity is conscious. Consciousness, therefore, is like tiny islands in an ocean of unconscious processing. Yet, while "jumping" from one island to the next through the abyss of the unconscious processes, our consciousness convinces our minds that we smoothly move through continually conscious states.

How the consciousness does it? At this point we would not dwell much on this, not enough yet is known. We would suggest that conscious perceptions of our mental states is due to special mental representation-model devoted to consciousness, within this model we can, to some extent, control our will and use it to direct our attention. The illusion of smooth consciousness is created by properties of this model. This illusion has a clear survival value: it would be difficult to survive with consciousness switched off 99.99% of time, and "lighting up" when individual objects or concepts come to consciousness for only 0.01% of time. It is known from many psychological experiments, that we actually do not perceive everything around us with equally conscious attention. Much of what we take for a smooth conscious perception of the surrounding world is "fillings in" for the blanks; these fillings in are based on expectations, on what we saw recently, and they are not as crisp and not as conscious as a single object in the center of our attention. An in-depth discussion of the properties of the "conscious model" is beyond the scope of this book, although we will dwell on some aspects of it later when discussing higher-level cognition, symbols, and interaction between language and cognition. We would emphasize that much ongoing discussions of consciousness are misdirected. There is no "Consciousness" with capital "C," consciousness is a matter of degree. Scientific goals in studying consciousness are similar to other scientific goals: being able to predict observed phenomena, identify mechanisms of the brain-mind correlated with subjective feelings of being conscious, understand relations between conscious perceptions and reality.

To summarize, the basic properties of perception and cognition processes, the working of the mental models, are well modeled mathematically by the DL process "from vague-to-crisp." The significant part of this process is unavailable to consciousness, vaguer perceptions are less conscious, and crisper are more conscious. These predictions of DL have been demonstrated in neuro-imaging experiments.

4.2.2 Imagination and Intuition

Imagination, as already mentioned, involves excitation of a neural pattern in a sensory cortex in absence of an actual sensory stimulation. For example, visual imagination involves excitation of visual cortex, say, with closed eyes [31,43,89]. Imagination was long considered a part of thinking processes; Kant [41] emphasized the role of imagination in the thought process, he called thinking "a play of cognitive functions of imagination and understanding," Whereas pattern recognition and artificial intelligence algorithms of recent past would not know how to relate to this [48,50], Carpenter and Grossberg's adaptive resonance model [12,29,30] and NMF both describe imagination as an inseparable part of thinking. Imagined patterns are top-down signals that prime the perception cortex areas (priming is a neural terminology for making neurons to be more readily excited). In NMF, the imagined neural patterns are given by models $\mathbf{M}_m$.

Visual imagination, as mentioned, can be "internally perceived" with closed eyes. The same process can be mathematically modeled at higher cognitive levels, where it involves models of complex situations or plans. Similarly, models of behavior at higher levels of the hierarchy can be activated without actually propagating their output signals down to actual muscle movements and to actual acts in the world. In other words, behavior can be imagined, along with its consequences, it can be evaluated, and this is the essence of plans. Sometimes, imagination involves detailed alternative courses of actions considered and evaluated consciously. Sometimes, imagination may involve fuzzy or vague, barely conscious models in the process of adaptation, which reach consciousness only after they converge to a "reasonable" course of action, which can be consciously evaluated. From a mathematical standpoint, this latter mechanism is the only possible; conscious evaluation cannot involve all possible courses of action; it would lead to combinatorial complexity and impasse. At lower levels of perception this has been demonstrated in neuroimaging experiments. Extending similar experimental demonstrations to higher cognitive levels is a challenge for the future research.

In agreement with neural data, the KI theory adds details to Kantian description: thinking is a play of top-down higher-hierarchical-level imagination and bottom-up lower-level understanding. Kant identified this "play" as a source of aesthetic emotions that we consider in the next section. Kant used the word "play," when he was uncertain about the exact mechanism; this mechanism is KI and NMF-DL.

Intuitions include inner perceptions of models, imaginations produced by them, and their relationships with objects and events in the world. Their mathematical-psychological status is similar to examples considered in chapter 3, say, Figs 3d through 3g; but the whole process of Fig.3 during visual perception is fast, it takes about 170 ms, and usually it does not reach consciousness until Fig.3h when it becomes conscious perception. What is subjectively perceived as intuition includes higher-level models of relationships among simpler models; while the higher-level models are in the process of their development, especially when this development takes a long time. Intuitions involve vague-fuzzy unconscious concept-models, which are in a state of being formed, learned, and being adapted toward crisp and conscious models (say, a theory). Conceptual contents of vague models are undifferentiated and partly unconscious. Below we discuss emotions related to understanding, here we just mention that similar to conceptual vagueness, conceptual and emotional contents of these vague mind states are undifferentiated; vague concepts and emotions are mixed up. Vague mind states may satisfy or dissatisfy the desire to understand in varying degrees before they become differentiated and accessible to consciousness, hence the vague complex emotional-cognitive feel of an intuition. Contents of intuitive states differ among people, but the main mechanism of intuition, according to NMF is the same among artists and scientists. Composers' intuitions are mostly about sounds and their relationships to psyche. Painters' intuitions are mostly about colors and shapes and their relationships to psyche. Writers' intuitions are about words, or more generally, about language and its relationships to psyche. Mathematicians' intuitions are about structure and consistency within a theory, and about relationships between the theory and a priori content of psyche. Physicists' intuitions are about the real world, first principles of its organization, and mathematics describing it. These suggestions are hypotheses that should be verified in psychological and neural experiments. Let me repeat that contents of this section is a summary of many publications discussed in the Literature section.

4.2.3 The Knowledge Instinct and Emotions

We continue describing relationships between DL and high-level mental abilities. Computational intelligence is closer to workings of the mind than it is commonly believed. Some discussions in this section have been psychologically established, others are hypotheses subject to experimental verification. A better understanding of the mind is offered here than is the current state of the art (although all discussions in this book, including this chapter, have been well published, they are scattered through many specialized journals in areas of mathematics, engineering, psychology, cognitive science, philosophy, aesthetics, and musicology, in several languages; and this chapter is unique in combining this diverse knowledge). Along with discussing mathematical models of the mind mechanisms, we briefly discuss psychological and neural experiments supporting the proposed theory. We also mention directions to proceed in order to verify this theory using psychological and neural experiments.

Discussions here continue a long tradition of attempts to understand workings of the mind, including Adaptive Resonance Theory (ART) that is related to NMF by emphasizing interactions of bottom-up and top-down signals . Understanding of the mind is necessary for engineering the next-level intelligent systems, including collaborative human-computer systems, which will understand language and experience emotions as part of their cognitive mechanisms. Some of these systems are already under development, as discussed throughout this book. Similarly, some psychological contents of this section are either known in neuro-psychology, or are the subject of ongoing experimental verifications. In addition, we outline future neuro-psychological experimental programs.

The functioning of the mind and brain cannot be understood in isolation from the system's "bodily needs," A biological system needs to replenish its energy resources (to eat). This and other fundamental unconditional needs are indicated to the system by instincts. Scientific terminology in this area is still evolving. E.g. psychologists prefer the word *drives* and avoid a word *instincts*. The reason is that historically instincts were mixed up with instinctual behavior and other less useful terms, mixing up fundamental mechanisms and complex behavior requiring explanations. A mathematical theory in this chapter describes neural mechanisms of the mind from the first principles, which are clearly defined. We describe instincts mathematically as internal sensors, which measurements directly indicate unconditional needs of an organism. Various instincts can be described in more details of the underlying neural mechanisms, however, for our purpose of describing the main mind mechanism, the above definition is sufficient. For example, instinct for food measures the sugar level in the blood. Our bodies have many internal sensors measuring body states essential for survival, such as blood pressure, temperature, etc.

How do instinctual measurements affect our thinking and behavior? Clearly, we do not consciously "read" instinctual sensor "dials," Instinctual needs are made available to decision-making parts of our brains by emotional neural signals Satisfactions of instinctual needs are felt as positive emotions, dissatisfaction as negative. In this way emotional signals affect processes of perception and cognition. Objects satisfying instinctual needs receive priority in perception and recognition. For example, when the sugar level in the blood gets low, we feel the corresponding emotional signals as hunger, and recognition of food objects receives priority over other objects.

In this chapter we concentrate in details on a single instinctual mechanism, the knowledge instinct, KI, which is described mathematically as a maximization of similarity measure between bottom-up and top-down signals, L, eq.(2.1-1). Without matching bottom-up and top-down signals perception will not function, and we will not be able to survive. Therefore KI is a most fundamental instinct, more fundamental than instincts for procreation or avoiding danger.

Biologists and psychologists have discussed various aspects of this mechanism, a need for positive stimulations, curiosity, a motive to reduce cognitive dissonance, a need for cognition. Until recently, however, this instinct or drive was not mentioned among 'basic instincts' on a par with instincts for food and procreation.

The fundamental nature of this mechanism became clear during mathematical modeling of workings of the mind. Our knowledge always has to be modified to fit the current situations. We don't usually see exactly the same objects as in the past: angles, illumination, and surrounding contexts are different. Therefore, our mental representations have to be modified; adaptation-learning is required. Virtually all learning and adaptive algorithms maximize correspondence between objects of recognition and an algorithm internal structure (knowledge in a wide sense); the psychological interpretation of this mechanism is KI. Below we discuss the mind-brain mechanisms of KI. As we discuss below, KI is a foundation of our higher cognitive abilities, and it defines the evolution of consciousness and cultures.

Satisfaction or dissatisfaction of KI is felt emotionally. What kind of emotion is related to knowledge? At lower levels of perception of minute details or everyday objects KI mechanisms function autonomously, below the level of consciousness, and emotions of KI satisfaction are not conscious. However, as soon as the autonomous functioning of KI fails, if we cannot recognize familiar surroundings, we immediately feel this dissatisfaction with emotions. This mechanism is a standard staple of thrillers, which show us situations that the mind cannot match to everyday mental models. From this example it is clear that satisfaction or dissatisfaction of KI is felt as harmony or disharmony between our knowledge and the surrounding world. At the level of everyday perception, autonomous KI functioning is similar to functioning of stomach; successful functioning occurs automatically, requires no attention, and we do not feel a strong emotion of harmony when our mental model of, say, a chair successfully matches the actual chair. However, as soon as stomach or perception do not function properly, we immediately feel it emotionally. At higher level of cognition, when successfully solving a problem that occupied us for a while, we feel emotionally a harmony between the problem and our solution.

Emotions of satisfaction or dissatisfaction of each bodily need, such as hunger or satiation, are called *prime* emotions; many have special words to describe them. Is there a special word for describing harmonious or disharmonious emotions related to KI? Since Kant these emotions related to knowledge and understanding are called *aesthetic* emotions. Later we describe how they relate to feelings of the beautiful and how they make up the foundations for all our higher mental faculties. Here we would like to emphasize that aesthetic emotions are not specific to artists or museums, they are inseparable from every act of perception and cognition.

4.2.4 Aesthetic Emotions and the Beautiful

In NMF, KI constantly generates emotional signals, which we perceive as feelings of harmony or disharmony between our knowledge and the world; these emotions drive us to improve our mind's models-concepts for better correspondence to surrounding objects and events.

Mathematically aesthetic emotions are given by changes in similarity measure dL/dt. When new data are coming, which do not correspond to existing models, the similarity change dL/dt is negative, understanding is low, and aesthetic emotions are negative, indicating dissatisfaction of the learning instinct. This stimulates learning. In the process of learning, dL/dt is positive, dynamic logic NMF emotionally enjoys learning. It might seem as an exaggeration, when we refer to a simple algorithm "enjoying" learning of simple patterns. However, when thousands of DL-NMF agents would understand the world (or Internet), while communicating among themselves and human users, the words "emotions" and "enjoy" would be more easy to accept as accurate description and similar to mechanisms of the human mind. We would emphasize that mechanisms of DL-NMF and KI given in chapters 2 and 3 is a step toward computational mechanisms of aesthetic emotions.

Cognitive science is at a complete loss when trying to explain the highest human abilities, the most important and cherished ability to create and perceive the beautiful. Its role in the working of the mind was not understood. Aesthetic emotions discussed above are often below the level of consciousness at lower levels of the mind hierarchy. Simple harmony is an elementary aesthetic emotion related to improvement of mental models of objects. Higher aesthetic emotions, according to NMF, are related to the development and improvement of more complex "higher" models at higher levels of the mind hierarchy. At higher levels, when understanding important concepts, aesthetic emotions reach consciousness.

Models at higher levels of the mind hierarchy are more general than lower-level models; they unify knowledge accumulated at lower levels. The highest forms of aesthetic emotions are related to the most general and most important models near the top of the mind hierarchy. According to Kantian analysis among the highest models are models of the meaning of our existence, of our purposiveness or intentionality. The hypothesis here is that KI drives us to develop these models. The reason is in the two sides of knowledge: on one hand knowledge consists in detailed models of objects and events required at every hierarchical level, on the other, knowledge is a more general and unified understanding of lower-level models at higher levels. These two sides of knowledge are related to viewing the knowledge hierarchy from bottom up or from top down; they are related to the mechanisms of bottom-up and top-down signals. In the top-down direction, models strive to differentiate into more and more detailed models accounting for every detail of the reality. In the bottom-up direction, models strive to make a larger sense of the detailed knowledge at lower levels. In the process of cultural evolution, higher, general models have been evolving with this purpose, to make more sense, to create more general meanings. In the following sections we consider mathematical models of this process of cultural evolution, in which top mental models evolve. The most general models, at the top of the hierarchy, unify all our knowledge and experience. The mind perceives them as the models of meaning and purpose of existence. In this way KI theory corresponds to Kantian analysis.

Everyday life gives us little evidence to develop models of meaning and purposiveness of our existence. People are dying every day and often from random causes. Nevertheless, belief in one's purpose is essential for concentrating will and for survival. Is it possible to understand psychological contents and mathematical structures of models of meanings and purpose of human life? It is a challenging problem yet NMF-DL gives a foundation for approaching it.

Let us remember the closed-eye experiment considered in section 4.1.1. Mental representations-models of everyday objects are vague, when we are not looking at these objects. We can conclude that models of abstract situations, higher in the hierarchy, which cannot be perceived with "opened eyes," are much vaguer. Even much vaguer have to be models of the purpose of life at the top of the hierarchy. As mentioned, everyday life gives us no evidence that such a meaning and purpose exist at all. And many people do not believe that life has a meaning. When I ask my colleagues-scientists, if life has a meaning, most protest against such a nebulous, indefinable, and seemingly unscientific idea. However, nobody would agree that his or her personal life is as meaningless as a piece of rock at a road wayside.

Is there a scientific way to resolve this contradiction? This is exactly what we intend to do in this chapter with the help of NMF-DL mathematical models and recent results of neuro-psychological experiments. Let us go back again to the closed eye experiment. Vague imaginations with closed eyes cannot be easily recollected when eyes are opened. Vague states of mental models are not easily accessible to consciousness. To imagine vague objects we should close eyes. Can we "close mental eyes" that enable cognition of abstract models? Later we consider mathematical models of this process. Here we formulate the conclusions. "Mental eyes" enabling cognition of abstract models involve language models of abstract ideas. These language models are results of millennia of cultural evolution. High-level abstract models are formulated crisply and consciously in language. To significant extent they are cultural constructs, and they are different in different cultures. Every individual creates cognitive models from his or her experience guided by cultural models stored in language. Whereas language models are crisp and conscious, cognitive models are vague and less conscious. Few individuals in rare moments of their lives can understand some aspects of reality beyond what has been understood in culture over millennia and formulated in language. In these moments "language eyes" are closed and an individual can see "imagined" cognitive images of reality not blinded by culturally received models. Rarely these cognitions better represent reality than cultural models developed over millennia. And even rarer these cognitions are formulated in language so powerfully that they are accepted by other people and become part of language and culture. This is the process of cultural evolution. We will discuss it in more details later.

Understanding the meaning and purpose of one's life has been important for survival millions of years ago and is important for achieving higher goals in contemporary life. Therefore all cultures and all languages forever have been formulating contents of these models. And the entire humankind has been evolving toward better understanding of the meaning and purpose of life. Those individuals and cultures that do not succeed are handicapped in survival and expansion. But let us set aside cultural evolution for later sections and return to how an individual perceives and feels his or her models of the highest meaning.

As we discussed, cognitive models at the very top of the mind hierarchy are vague and unconscious. Even so many people are versatile in talking about these models, and many books have been written about them, cognitive models that correspond to the reality of life are vague and unconscious. Some people, at some points in their life, may believe that their life purpose is finite and concrete, for example to make a lot of money, or build a loving family and bring up good children. These crisp models of purpose are cultural models, formulated in language. Usually they are aimed at satisfying powerful instincts, but not the knowledge instinct and they do not reflect the highest human aspirations. Reasons for this perceived contradiction are related to interaction between cognition and language that we have mentioned and will be discussing in more details later [55,62,68]. Everyone who has achieved a finite goal of making money or raising good children knows that this is not the end of his or her aspirations. The psychological reason is that everyone has an ineffable feeling of partaking in the infinite, while at the same time knowing that one's material existence is finite. This contradiction cannot be resolved. For this reason cognitive models of our purpose and meaning cannot be made crisp and conscious, they will forever remain vague, fuzzy, and mostly unconscious.

As discussed, improvement of models leads to better understanding of what the model is about, to satisfaction of KI, and to corresponding aesthetic emotions. Higher in the hierarchy models are vague, less conscious and emotional contents of mental states are less separated from their conceptual contents. At the top of the mind hierarchy, the conceptual and emotional contents of cognitive models of the meaning of life are not separated. In those rare moments when one improves these models, improves understanding of the meaning of one's life, or even feels assured that the life has meaning, he or she feels emotions of the beautiful, the aesthetic emotion related to satisfaction of KI at the highest levels.

These issues are not new; philosophers and theologians expounded them from time immemorial. The NMF-DL and knowledge instinct theory gives us a scientific approach to the eternal quest for the meaning. We perceive an object or a situation as beautiful, when it stimulates improvement of the highest models of meaning. Beautiful is what "reminds" us of our purposiveness. This is true about perception of beauty in a flower or in an art object. Just an example, R. Buckminster Fuller, an architect, best known for inventing the geodesic dome wrote: "When I'm working on a problem, I never think about beauty. I think only how to solve the problem. But when I have finished, if the solution is not beautiful, I know it is wrong". Similar things were told about scientific theories by Einstein and Poincare, emphasizing that the first proof of a scientific theory is its beauty. The KI theory explanation of the nature of the beautiful helps understanding an exact meaning of these statements and resolves a number of mysteries and contradictions in contemporary aesthetics.

Finishing scientific discussion of the beautiful, I would like to emphasize again that it is an emotion related to knowledge at the top of the mind hierarchy, the knowledge of the life meaning. It is governed by KI, not by sex and instinct for procreation. Sexual instinct is among the strongest of our bodily instincts, and it makes use of all our abilities, including knowledge and beauty. And yet the ability for feeling and creating the beautiful is related not to sexual instinct but to the instinct for knowledge.

4.3 Natural Language Learning

We briefly summarize directions in computational linguistics: Chomsky or nativist, cognitive, and evolutionary linguistics. Discuss that their computational failures are related to combinatorial complexity and logic. Then we discuss how DL for situation learning can be used to overcome this difficulty and to combine cognitive and evolutionary linguistics. We discuss an application to Internet search engines with elements of language learning.

4.3.1 Linguistic Theories Since the 1950s

Complex innate mechanisms of the mind were not appreciated in the first half of the last century. The thinking of mathematicians and the intuitions of psychologists and linguists were dominated by logic. Language and cognition, when understood as logical mechanisms seemed not much different; both were based on logical statements and rules.

Contemporary linguistic interests in the mind mechanisms of language were initiated in the 1950s by Chomsky (1965). He identified the first mysteries about language that science had to resolve. "Poverty of stimulus" addressed the fact that the tremendous amount of knowledge needed to speak and understand language is learned by every child around the world even in the absence of formal training. Compare learning language to learning quantum physics or theory of relativity. Physical theories are based on few first principles; the rest is deduced with few basic rules. Language, on the opposite, requires remembering tens of thousands of words and dozens of rules that every child learns effortlessly by the age of 5. How come that equal proficiency in physical theories is attained by much fewer people after years of education and practice. One way to understand it is that language learning is based on specific inborn mechanisms.

Chomsky has thought obvious that surrounding language cultures do not carry enough information for a child to learn language, unless specific language learning mechanisms are inborn in the mind of every human been. This inborn mechanism should be specific enough for learning complex language grammars and still flexible enough so that a child of any ethnicity from any part of the world would learn whichever language is spoken around, even if he or she is raised on the other side of the Globe. This inborn learning mechanism Chomsky called Universal Grammar and set out to discover its mechanisms. He emphasized the importance of syntax and thought that language learning is independent of cognition. This approach to language based on innate mechanisms, is called *nativism*.

Chomsky and his school initially used available mathematics of logical rules, similar to rule systems of artificial intelligence. In 1981 Chomsky proposed a new mathematical paradigm in linguistics, *rules and parameters*. This was similar to model-based systems emerging in mathematical studies of cognition (which we discussed in chapter 1). Universal properties of language grammars were supposed to be modeled by parametric rules or models, and specific characteristics of grammar of a particular language were fixed by parameters, which every kid could learn when exposed to the surrounding language. Another fundamental change of Chomsky's ideas (1995) was called the *minimalist program*. It aimed at simplifying the rule structure of the mind mechanism of language. Language was considered to be in closer interactions to other mind mechanisms, closer to the meaning, but stopped at an interface between language and meaning. Chomsky's linguistics still assumes that meanings appear independently from language. Logic is the main mathematical modeling mechanism.

Many linguists disagreed with the separation between language and cognition in Chomsky's theories. Cognitive linguistics emerged in the 1970s to unify language and cognition, and explain the creation of meanings. Cognitive linguistics rejected Chomsky's idea about a special module in the mind devoted to language. The knowledge of language is no different from the rest of cognition, and is based on conceptual mechanisms. It is embodied and situated in the environment. Related research on construction grammar argues that language is not compositional, not all phrases are constructed from words using the same syntax rules and maintaining the same meanings; metaphors are good examples (Croft & Cruse 2004; Evans & Green 2006; Ungerer & Schmid 2006). Feldman (2010) argues that by combining linguistic and cognitive constructions, construction grammar can explain both language and cognition as compositional. Cognitive linguistics so far has not lead to a computational theory of language or cognition, explaining how meanings are created. The formal apparatus of cognitive linguistics is dominated by logic and succumbs to CC.

Evolutionary linguistics emphasized that language evolved together with meanings. A fundamental property of language is that it is transferred from generation to generation, and language mechanisms are shaped by this process. (Hurford 2008; Christiansen & Kirby 2003). Evolutionary linguistics by simulating societies of communicating agents (Brighton, Smith & Kirby 2005; Fontanari & Perlovsky 2007) demonstrated the emergence of a compositional language. Yet, existing examples encountered combinatorial complexity and cannot be extended to realistic complexity of language.

4.3.2 DL for Learning Language

Combinatorial complexity encountered by computational approaches to language is related to the inherent combinatorics of language. Words are combinations of sounds. There are approximately 40 different sounds (phonemes) in English. Considering combinations of 7-phonemes or shorter, there is a possibility of forming almost a trillion words. Yet most of English speakers use less than few tens of thousands of words, 0.00001% of all possibilities. Even "worse" is the case of forming phrases from words. Considering combinations of 10,000 words into phrases of no longer than 7 words yields 10^{35} possibilities.

The mechanism of learning actual words of a language among all possible combinations of sounds contemporary linguistics is considering a straightforward remembering, plus few combinatorial morphological rules, such as "s" ending for plurals, and "ed" for past tense. Remembering several tens of thousands words is computationally possible and not too difficult. However, learning which word corresponds to which object is not possible by remembering alone: the number of combinations between ten thousand words and objects (or actions) is on the order of $10{,}000^{10{,}000}$, a number much larger than the Universe. No computer or brain would be able to make that many computations. (The idea that words and objects are associated through direct remembering is called "associationsim"; it is an old idea, which descends from Locke. Most psychologists today are unaware that this is mathematically untenable, and still think that associationsim is the mechanism combining language and cognition).

The learning of relations between words and objects or events is computationally possible by DL, by using a technique similar to the one used in section 3.7 for situation learning. Word-object relations can be learned from chunks of continuous speech and perceptions of objects; in this case situations include not only objects, but also chunks of speech. Similar, phrases are sets of words along with relations among words. As discussed in section 3.7, for this purpose, relations among words and the corresponding markers, indicating related words, are treated the same way as words. This way syntax is learned the same way as the rest of language. Syntax relates words in a phrase in correspondence with how objects and events are related in the world. Phrase structures (syntax) therefore have to be learned jointly with situations in the world and relations between objects, actions, and events.

This discussion along with the DL technique described in section 3.7 outlines how to overcome the principled difficulty of combinatorial complexity. Still, the engineering development of intelligent agents that understand language and the world along with relations between them is a problem to be addressed in future. Several aspects of this problem are addressed in the following sections.

When learning language, the knowledge instinct, which drives maximization of similarities between language models and language data, is called the language instinct, LI. The principled difference between KI and LI is that LI is only "concerned" with learning language models (words, phrases, symtax…), but does not match them to real-world objects, situations, actions, or experiences. LI is responsible for language, KI is responsible for cognition. This separation is responsible for separate learning of language (by the age of 5) and learning to understand the world (cognition, which goes on much longer). Integration of language and cognition is considered later.

4.3.3 Search Engines for the Internet with Elements of Learning Understanding

An engineering application of the above discussion could be search engines with language understanding. A first step toward this can be made by using simplified models of phrases, called bag models. Bag models ignore syntax and grammar and consider phrases just as sets of words without any relations. This simplified language can be learned directly by using section 3.7 technique. What we have called objects in section 3.7 are words, and situations are sets of words, or bag models of phrases.

The straightforward simplified development could (1) identify most frequently used phrases, (2) characterize every document by phrases, and (3) develop a user interface that would identify phrases corresponding to a query. If a simple search is unsatisfactory, a user would select appropriate phrases suggested based on the query, and search by phrases, rather than by words. Using bag-models would miss part of real language information, still would be much better than search by key-words.

The next step could be to develop several hierarchical levels. A level above phrases could use "bags of phrases" models, etc. Essential step in this development is an efficient user interface. The principled problem of combinatorial complexity of language is solved.

Next steps could introduce elements of syntax.

4.4 Integration of Language and Cognition

4.4.1 Language and Cognition

Do we use phrases to label situations that we already have understood, or the other way around, do we just talk without understanding any cognitive meanings? It is obvious that different people have different cognitive and language abilities and may tend to different poles in the cognitive-language continuum, while most people are somewhere in the middle in using cognition to help with language, and vice versa. What are the computational mechanisms avoiding combinatorial complexity and the neural mechanisms that enable this flexibility? How do we learn which words and objects come together? If there is no specific language module, which is assumed by Chomsky's linguists and rejected by cognitive linguists, why do kids learn a language by 5 or 7, but do not think like adults?

Little is known about neural mechanisms for integrating language and cognition. This section outlines a computational model that potentially can answer the above questions, and that is computationally tractable, it does not lead to combinatorial complexity and can be used for engineering applications. Also it implies relatively simple neural mechanisms, explains why human language and human cognition are inextricably linked, and possibly sets us aside from animals. It suggests that human language and cognition have evolved jointly.

4.4.2 Dual Model

Whereas Chomskyan linguists could not explain how language and cognition interact, cognitive linguists could not explain why kids learn language by 5 but cannot think like adults; neither theory can overcome combinatorial complexity. We propose that the integration of language and cognition is accomplished by the dual model. Every concept-model $\mathbf{M}_m$ has two parts, linguistic $\mathbf{M}^L{}_m$ and cognitive $\mathbf{M}^C{}_m$:

$$\mathbf{M}_m = \{ \mathbf{M}^C{}_m , \mathbf{M}^L{}_m \}; \qquad (4.4.1)$$

As a sensor data stream constantly comes into the mind from all sensory perceptions; every part of this data stream is constantly evaluated and associated with models according to the mechanisms of dynamic logic described in section 3.7. In a newborn mind both types of models are vague and mostly empty placeholders for future cognitive and language contents. The neural connections between the two types of models are inborn; the mind never has to learn which word goes with which object. As models acquire specific contents in the process of growing up and learning, linguistic and cognitive contents are always staying properly connected.

During the first year, infants learn some objects and situations in the surrounding world. In terms of DL, this means that cognitive parts of some models at the level of objects and situations become less vague and acquire a degree of specificity. Language models at the level of objects and above remain vague. After one year of age, language model adaptation speeds up; language models become less vague and more specific much faster than the corresponding cognitive models. This is especially true about contents of abstract models, which cannot be directly perceived by senses, such as "law," "state," "rationality." This explains how it is possible that kids by the age of five can talk about most of contents of the surrounding culture but cannot function like adults: language models are acquired ready-made from the surrounding language, but cognitive models remain vague and gradually acquire concrete contents throughout life by accumulating cognitive models in correspondence with language models. This is the neural mechanism of what is colloquially called "acquiring experience."

Human learning of cognitive models continues through the lifetime and is guided by language models. The knowledge instinct drives the human mind to develop more specific and concrete cognitive models by accumulating experience throughout life in correspondence with language models. Language learning as discussed is driven by a somewhat different mechanism of LI. Language learning is grounded in the surrounding language, which has accumulated cultural wisdom in ready-made language models. This is the reason why language can be learned much earlier than the real world understanding, cognition. We will repeat that language is learned ready-made, it accumulates millennia of cultural knowledge, but cognition requires personal life experience, it accumulates slowly. The dual model enables this two-step learning.

4.4.3 Experimental Evidence, Answers and Questions

Michael Arbib (2005) suggested that parts of the brain involved in language evolved on top of the system of mirror neurons. Mirror neurons in humans, primates, and some birds are specialized neurons activated both, when an animal performs an action and when it observes the same action performed by another animal. In primates, mirror neurons are located in the Broca's area of the brain, which is also associated with language in humans. So, connections between language brain areas and perception of actions and events, which are required for the dual model, have evolved long before language. Evolution "prewired" the human brain for language.

Experimental evidence for the dual model began to emerge. The first experimental indication has appeared in (Franklin et al 2008). They have demonstrated that categorical perception of color (say, category of blue vs. category of green) in prelinguistic infants is based in the right brain hemisphere. As language is acquired and access to lexical color codes (words for color) becomes more automatic, categorical perception of color moves to the left hemisphere (between two and five years) and adult's categorical perception of color is only based in the left hemisphere (closer to language).

These experiments have provided evidence for neural connections between perception and language, a foundation of the dual model. Possibly it confirms another aspect of the dual model: the crisp and conscious language part of the model hides from our consciousness the vaguer cognitive part of the model. This is similar to what we observed in the close-open eye experiment: with opened eyes we are not conscious about vague imaginations-priming signals.

So, we can answer some of the questions posed at the beginning of the section. Language and cognition are separate and closely related mechanisms of the mind. They evolve jointly in ontological development and learning, and likely these abilities evolved jointly in evolution—this we address in more details in the next section. This joint evolution of dual models from vague to more crisp content resolves the puzzle of associationism: there is no need to learn correct associations among combinatorially large number of possible associations, words and objects are associated all the time while their concrete contents emerge in the mind.

Perception of the objects that can be directly perceived by sensing might be to some extent independent from language, nevertheless, as the above experimental data testify, even in these cases language affects what we perceive. In more complex cognition of abstract ideas, which cannot be directly perceived by senses, we conclude that the language parts of the models are more crisp and conscious; language models guide the development of the content of cognitive models. Language models also tend to hide the vaguer cognitive contents from consciousness. It follows that in everyday life most thinking at higher abstract levels is accomplished by using language models, possibly with little engagement of cognitive contents.

We know that thinking by using cognitive contents is possible, for example when playing chess. Mathematics is another example, but not necessarily a good one, because mathematics uses its own "language" of mathematical notations. Do mathematical notations play a similar role to language in shadowing cognitive contents and thinking? Our guess would be that this is the case for a C student of mathematics, but creative thinking in mathematics and in any other endeavor engages cognitive models. Needless to say, this requires special abilities and significant effort. Possibly a brain region fusiform gyrus plays a role in cognition shadowed by language. More detailed discussion of possible brain regions involved in the knowledge instinct are discussed in (Levine & Perlovsky 2008). This is a vast field for experimental psychological and neuro-imaging investigations.

4.4.4 Dual Hierarchy

The dual model implies two parallel hierarchies of language and cognition, as illustrated in Fig.4.4-1. This architecture along with the DL equations in section 3.7 solve an amazing mystery of the human mind, which we are so used to that it almost never has been even formulated as requiring an explanation.

The cognitive models at higher levels are composed of lower level models (say, situations are composed of objects, as in section 3.7; more accurately to say, higher level models are composed of patterns of bottom-up signals from lower level models). In parallel, language is used to describe the situations linguistically with phrases composed of words. Words-object relations at the lower levels are preserved at higher levels of phrase-situation relations. This holds true across a number of phrase-situation level models, using various combinations of the same words from the lower level. This amazing property of our mind seems so obvious, that the nontrivial complexity of the required mechanism has only been noticed once (Deacon, 1997). The dual hierarchy explains this by suggesting that learning of higher-level cognitive models is guided by language models at the same level.

Deacon also suggested that the hierarchy sets the human mind apart from the animal world. Here we discuss the mathematical reasons why hierarchy can only exist as a joint dual hierarchy of language and cognition. Every human culture possesses both abilities; and there is no species that has either language or cognition at the human level. This dual hierarchy architecture gives a mathematical reason for this fact. Only at the lower levels in the hierarchy can cognitive models be learned by direct perception of the world. Learning is grounded in "real" objects. At higher levels, however, learning of cognitive models has no ground in the world. In artificial intelligence it was long recognized that learning without grounding could easily go wrong, learned or invented models may correspond to nothing real or useful (Meystel and Albus 2001). In section 3.7 we demonstrated learning of cognitive models of situations from a limited number of examples. The fundamental problem solved there was the overcoming of

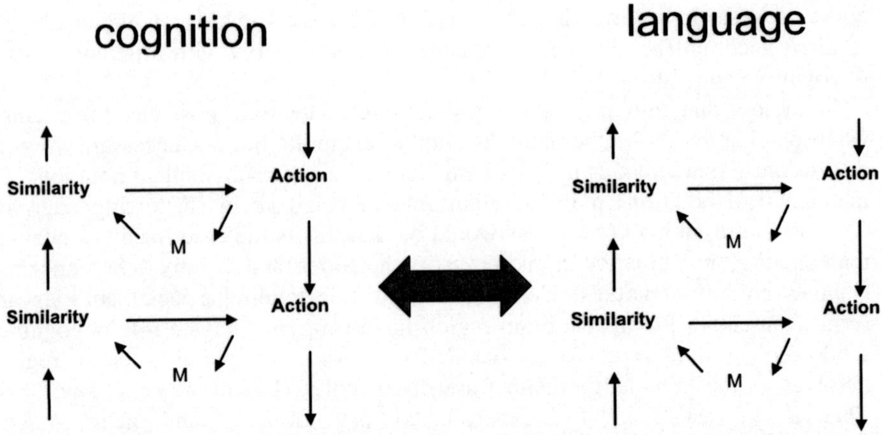

Fig. 4.4-1 Hierarchical integrated language-cognition NMF system. At each level in a hierarchy there are integrated language and cognition models. Similarities are integrated as products of language and cognition similarities. Initial models are fuzzy placeholders, so integration of language and cognition is sub-conscious. Association variables depend on both language and cognitive models and signals. Therefore language models help cognitive model learning. High-level abstract cognitive concepts are grounded in abstract language concepts, which in turn are grounded in the surrounding language at all levels. Initial models are fuzzy placeholders, so integration of language and cognition is sub-conscious. Association variables depend on both language and cognitive models and signals. Therefore language model learning helps cognitive model learning (and v.v. at lower levels). Learning language is driven by the language instinct and cognitive learning is driven by the knowledge instinct.

combinatorial complexity. However, separating useful situations from random one must have its limits: as the number of random, irrelevant combinations of lower-level models becomes larger, *grounding* is needed to learn useful models rather than random. At higher levels of abstract models, as mentioned, grounding cannot be based in experience. This is why learning of high-level cognitive models must be grounded in language. Language, in turn, is grounded in experience (in using language) at all hierarchical levels. This is illustrated in Fig. 4.4-2

The mechanism of the dual model sets the human mind apart from the rest of the animal world. Consider an example of a dog learning to bring shoes to a human master on a verbal command. A dog, it seems, can jointly learn language and cognition (a word "shoes" and an object shoes), does it mean that the dog possesses a dual model? No. The dog can do it, because it perceives an object (shoes) in the world. Learning a word "shoes" is grounded in direct perception of object-sound "shoes." Such a direct grounding in sensory signals exists only at the very bottom of the mind hierarchy. At higher levels, as mentioned, cognitive concepts are grounded in language concepts due to the dual models. Using the dual models, the knowledge instinct drives the mind to acquire cognitive models corresponding to language models (colloquially, "experience").

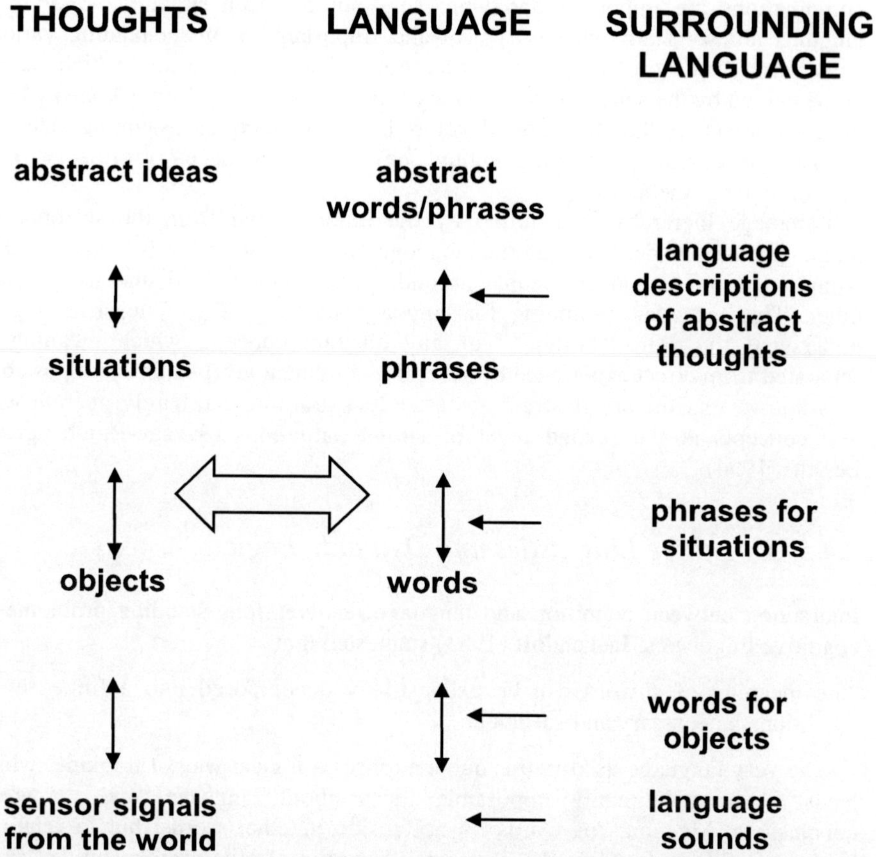

Fig. 4.4-2 Lower level cognitive models (on the left) are grounded in direct experience. At higher levels of abstract models, grounding cannot be based in experience. This is why learning of high-level cognitive models must be grounded in language. Language, in turn, is grounded in experience (in using language) at all hierarchical levels.

The fact that the cognitive hierarchy cannot be learned without language hierarchy is so fundamental and underappreciated that I would give another explanation for this reason in different words. Consider learning *situations* on top of already learned *object* perception. When deciding which set of objects constitutes a concept-model that should be learned and remembered, one would encounter a situation such as follows: entering a room one sees a desk, a chair, books, shelves… and a barely visible scratch on the wall. Is this scratch on the wall as important as other objects? Is the fact that a red-color book is on the left from a blue-color one important and should it be learned as a separate situation-model? In fact there are many more insignificant objects and infinity of their combinations, such as scratches, dust, relative positions, etc., than objects and their combinations that are significant for every situation. No one would have enough experience in a lifetime to learn which of the objects and in which combinations are important and which are not for each situation. Only from language do we learn what is typical and important for understanding various situations. Even more complicated is learning of abstract concepts, which cannot be perceived by the senses directly. Many linguists beginning from Chomsky have under-appreciated this fact and develop linguistic theories assuming that the theories of cognitive meaning should be handed down by professors from cognitive departments.

Language hierarchy is acquired by the human mind from the surrounding language ready-made. Learning the language hierarchy at all levels is grounded in communication with other people around; people talk to and understand each other. This provides grounding for language learning. Try to teach a dog to understand the word "rational," or any abstract concept, which meaning is separated from direct experience by several hierarchical levels; this is not possible. It is known that the smartest chimps after long training can barely operate with few concepts at the second level of simple situations (Savage-Rumbaugh & Lewine, 1994).

4.4.5 Cognitive Linguistics and Dynamic Logic

Interaction between cognition and language resolves long-standing problems of cognitive linguistics. Jackendoff (1983) suggested that

"the meaning of a word can be exhaustively decomposed into a finite set of conditions... necessary and sufficient..."

The very language used in this quote exposes a logical way of thinking, which leads to computationally impossible ideas about language, and to wrong conclusions. Meanings of words do not reside in other words, but in relations between words and real world situations. According to the mechanism described above, meanings reside in the cognitive parts of the dual models.

Gradually, cognitive linguistics moved away from the strictly compositional view of language. Lakoff (1988) emphasized that abstract concepts used by the mind for understanding the world have a metaphorical structure. Metaphors are not just poetic tools, but a mind mechanism for creating new abstract meanings. Lakoff's analysis brought this cultural knowledge of the role of metaphorical thinking within the mainstream of science. There was still a big gap between Lakoff's analysis of metaphors on one hand and neural and mathematical mechanisms on the other. The "Metaphors we live by" is a metaphorical book (the pun is intended) in that it begs the question: Who is that homunculus in the mind, interpreting the metaphorical theater of the mind? What are the mechanisms of metaphorical thinking? According to the current section, a metaphor extends an old understanding to the new meaning in a bidirectional interaction between language and cognition. A vague cognitive model is extended to a crisp language model by first making it vaguer (a metaphor); this is followed by the dynamic logic creation of several more specific new language and cognitive models.

In the works of Jackendoff (1983), Langacker (1988), Talmy (1988) and other cognitive linguists it was recognized that dichotomies of meanings (semantic-pragmatic) and dichotomies of hierarchical structures (superordinate-subordinate) were limiting the scientific discourse and have to be overcome. Consider the following opinions on creating meaning:

"in a hierarchical structure of meaning determination the superordinate concept is a necessary condition for the subordinate one... COLOR is a necessary condition for determining the meaning of RED" (Jackendoff, 1983).

"The base of predication is nothing more than... domains which the prediction actually invokes and requires" (Langacker, 1988)

These judgments illustrate the difficulties encountered when attempting to overcome old dichotomies. Logical intuitions guide these judgments and limit their usefulness. Attempts to implement mathematically the mechanisms assumed by these examples would lead to combinatorial complexity. Problems of meaning and hierarchy still reminded the old question about the chicken and the egg, what came first? If superordinate concepts come before subordinate ones, where do they come from? Are we born with the concept COLOR in our minds? If predications invoke domains, where do domains come from? These complex questions with millennial pedigrees are answered mathematically in this section. Hierarchy and meaning are emerging jointly with cognition and language. In evolution and individual learning, superordinate concepts (COLOR) are vaguer, less specific, and less conscious than subordinate ones (RED). RED can be vividly perceived, but COLOR can not be perceived. RED can be perceived by animals. But, the concept COLOR can only emerge in the human mind, due to joint operation of language and cognition via dual model.

Jackendoff in his recent research (2002) concentrated on unifying language and cognition. He developed detailed models for such unification; however, his logical structures face combinatorial complexity.

Lakoff and Johnson (1999) brought within the realm of linguistics an emphasis on the embodiment of the mind. They implied that in view of their discussions the entire philosophical tradition will have to be reassessed; this however, is an exaggeration. Recent synthesis of the computational, cognitive, neural, and philosophical theories of the mind demonstrated the opposite (Perlovsky, 2001). Plato, Aristotle, and Kant, even in specific details about the mind mechanisms, were closer to contemporary computational theories, than the 20th c. philosophers and mathematicians developing logical formalism and positivism.

Talmy (2000) introduced a notion of open and closed classes of linguistic forms. The open class includes most words, which could be added to language as needed, say, by borrowing from other languages. The closed class includes most grammatical structures (such as a, the, I, he, she…), which are fixed for generations and cannot be easily borrowed from other languages. This pointed to an important aspect of the interaction between language and cognition. Forms of the closed class interact with cognitive concepts, which emerged over thousands of years of cultural and language evolution. Thus, for each individual mind and for entire generations, which operate within the constraints of existing grammar, many cognitive concepts are predetermined by language. Talmy identified cognitive concepts affected by closed forms. These forms are more basic for cognition than words and unconsciously influence entire cultures. The idea of the closed class does not imply that certain cognitions of one culture cannot be understood in another culture, speaking a different language. For everyday cognitive constructs there are adequate expressions in many languages; however, interactions among multiple language and cognitive constructs might be different. These differences accumulate up the hierarchy, where cultural constructs are farther removed from direct experience. This creates differences among cultures, which origins and depths may not be obvious.

Kay (2002) proposed construction grammar (a direction closely associated with cognitive linguistics) to accommodate metaphoric and idiomatic linguistic constructions. These constructions reject the word-phrase distinction and adopt a word-phrase continuum, which is of course impossible as it would lead to combinatorial complexity. These constructions require a combination of semantic and linguistic knowledge. Yet, unlike the dual model and DL, existing proposals for construction grammar are logical and combinatorially complex; they do not explain how words are related to the world computationally. The dual model, instead, provides a necessary mechanism—cognitive and linguistic models act jointly.

Another proposal to combine cognition and language is given by Fauconnier & Turner (2008). It's intended function is similar to the dual hierarchy. They proposed what they called a double-scope blending, without however a specific computational mechanism to accomplish this.

Dual hierarchy supports the cognitive linguistic idea that syntax is not a separate inborn "box" in the mind, but is a conceptual mechanism. Mathematically, this suggestion is similar to what we discussed in section 3.7 about relations among objects. Relations are important for understanding situations. In section 3.7 we suggested that a relation can be described mathematically similar to an object. In addition, a relation requires specifying which objects are related; this is described by markers, which are also similar to objects and learned in a similar way. Syntactic relations among words can be modeled by a similar computational mechanism: by relations and markers. In English relations are specified by the word order, prepositions and other special words. In some other languages (including Mid-English and Old English) this function is also served by grammatical cases (nouns, pronouns, and adjectives may change to indicate their relations to other words).

Other specifics of syntax may be encoded in contents of concept-models at higher levels of phrases and sentences. We suggest that given a mechanism of the dual model, the hierarchy would evolve and the syntax would be learned from surrounding language. To which extent the syntax reflects structures in the world that could be directly learned along with language and encoded in cognitive and language models? What determines syntactic differences among languages: random historical events, or conditions of the environment? In addition to the dual model, what other linguistic knowledge must be inborn? What is the role of dynamic logic and the dual model in morphology (structure of words)? It is possible that the dual model is the only mechanism that is required to enable language and that sets us aside from animals. The described computational mechanisms of the dual model and DL make possible to address these questions in simulations and develop corresponding engineering systems. These are challenges for future research.

4.4.6 Evolutionary Linguistics and Dynamic Logic

Evolutionary linguistics emphasizes that language properties evolved in the process of the cultural evolution of languages (Christiansen and Kirby 2003). Only those properties of languages survive that can be passed from generation to generation. Christiansen and Chater (2008) discussed how various linguistic phenomena are explained within the evolutionary framework. Brighton et al (2005) demonstrated in a mathematical simulation that the evolutionary approach can explain the emergence of language compositionality. Compositionality, the language ability to construct words of sounds, phrases of words etc., is a fundamental language universal, unique to human languages. The Brighton et al work is especially elegant in that simple assumptions were required, first an ability to learn statistically from limited examples which sounds go with which cognitive meanings, and second, the fact that training samples are insufficient and agents have to guess the sounds for new meanings; so meanings that are similar in a certain way to the old ones are designated by sounds similar in some ways to the old sounds. This has led to compositionality.

A similar idea was demonstrated in (Fontanari & Perlovsky, 2007). They considered two conditions of a realistic language emergence. First, random errors in communications among speakers. And second, gradual assimilation of meanings, which have to occur in the evolution of a language, when new meanings are created by variations of old meanings. This case leads to the emergence of a language that is first, compositional, and second, preserves neighborhood relationships. That is relationships that map similar signals into similar meanings. So, these two important language properties naturally emerge in a realistic setting of language evolution.

Unfortunately, most work in evolutionary linguistics so far assumed that cognitive meanings already exist. This is unrealistic, as we have argued, at higher levels of the hierarchy of abstract cognition, language guides cognition and not vice versa. Another detrimental aspect of the existing work is the logical computational basis, leading to combinatorial complexity. Dynamic logic, if applied to Brighton et al's formulation, or to Fontanari & Perlovsky leads to non-combinatorial complexity of learning and production. If combined with the dual model, it does not require an assumption that the meanings have already existed; a result is the joint learning of combinatorial language and cognitive meanings (Fontanari, J. F., Tikhanoff, V., Cangelosi A. & Perlovsky, L. I. 2009). The mathematical formulation in this chapter leads to a joint evolution of language and cognitive meanings.

4.4.7 Contents of Language Faculty

A fundamental aspect of Chomsky's linguistics is language faculty, an inborn mechanism of language learning (in linguistics, the fundamental difficulty of explaining how this might happen is emphasized by using word *acquisition*). What exactly is the inborn content of this mechanism? Above we emphasized that DL and the dual model might be sufficient. Here we would like to compare our hypothesis to several influential concepts of Chomsky's linguistics. Hauser, Chomsky, and Fitch (2002) emphasized that "language is, fundamentally, a system of sound-meaning connections." This connection is accomplished by a language faculty, which generates internal representations and maps them into the sensory-motor interface, and into the conceptual-intentional interface. In this way sound and meaning are connected. Let us repeat, this book emphasizes that this assumption of a separate evolution of language from the sensory-motor and conceptual-intentional mechanisms would unavoidably lead to physically and biologically unrealizable combinatorial complexity of the learning combinations among separately evolved entities. In this subsection we address in some details mechanisms proposed in the above reference, and how they could be better performed by mechanisms of the dual model and DL.

Hauser, Chomsky, and Fitch (2002) emphasized that the most important property of the language faculty is recursion. However, they did not propose specific computational mechanisms how recursion creates representations, or how it maps representations into the sensory-motor or conceptual-intentional interfaces.

A conclusion of the previous discussions in this chapter is that it might not be necessary to postulate recursion as a fundamental property of a language faculty. In terms of the computational model of NMF-DL and the dual model proposed in this book, recursion is accomplished by the hierarchy: a higher level generates the next lower level models, etc., this accomplishes recursive functions. We have demonstrated that the dual model is a necessary condition for the hierarchy of language-cognition representations. It also might be a sufficient one. Although, this hypothesis has to be proven in future research, later we address the corresponding mechanisms. It is expected that the hierarchy is not a separate inborn mechanism; the hierarchy might emerge in operations of the dual model and dynamic logic in a society of interacting agents with intergenerational communications. What inborn precursors are necessary for the hierarchy ontological emergence, if any, is a challenge for the ongoing research.

By reformulating the property of recursion in terms of a hierarchy, along with demonstrating that a hierarchy requires the dual model, this chapter has suggested a new explanation: a single neurally-simple mechanism is unique for human language and cognitive abilities. Initial experimental evidence indicates a support for the dual model, still further experiments elucidating properties of the dual model are needed.

Another conclusion of this chapter is that the mechanism mapping between linguistic and cognitive representations is accomplished by the dual models. In previous sections we considered the mathematical modeling of the "conceptual-intentional interface" for intentionality given by the knowledge and language instincts; in other words we considered only intentionalities related to language and knowledge. It would not be principally difficult to add other types of intentional drives following (Levine and Grossberg 1987). The current book has not considered the "sensory-motor interface," which of course is essential for language production and hearing. This can be accomplished by the same mechanism of the dual model, with addition of behavioral and sensorial models. This task is not trivial; still it does not present principal mathematical difficulties.

We would also like to challenge an established view that specific vocalization is "arbitrary in terms of its association with a particular context." In animals, voice directly affects ancient emotional centers. In humans these affects are obvious in songs, and still persist in language to a certain extent. It follows that the sound-intentional interface is different from the language-cognition interface modeled by the dual model. The dual model frees language from emotional encumbrances and enables abstract cognitive development to some extent independent from primitive ancient emotions. Arbitrariness of vocalization (even to some extent) could only be a result of long evolution of vocalizations from primordial sounds (Perlovsky 2007).

Following sections consider remnants of primordial emotionality of sounds in languages. In evolution of languages, the genetic inborn fusion of emotions and concepts was replaced by habitual relations. We consider their mechanisms, and their necessity. If emotional-conceptual connections disappear, words without emotionality lose meanings; the entire language loses intentionality and meaning. This affects the entire cultural evolution. Following sections will touch on mathematical modeling of these effects. Yet detailed understanding of evolutionary separation of cognition from direct emotional-motivational control and from immediate behavioral connections remain challenges for future research.

4.4.8 Experimental Evidence and Future Research

The proposed mechanism of the dual model implies a minimal neural change from the animal to the human mind: it corresponds to Arbib's hypothesis about "language prewired" brain (2005) discussed later. It has emerged through combined cultural and genetic evolution, and cultural evolution continues today. DL and the dual model resolve a long-standing mystery of how human language, thinking, and culture could have evolved in a seemingly single big step, too large for an evolutionary mutation, too fast and involving too many advances in language, thinking, and culture, happening almost momentarily around 50,000 years ago (Deacon 1997; Mithen 1998). DL along with the dual model explain how changes, which seem to involve improbable steps according to logical intuition, actually occur through continuous dynamics. The developed theory provides a mathematical basis for modeling the concurrent emergence of hierarchical human language and cognition.

Evolutionary linguistics and cognitive science have to face a challenge of studying and documenting how the primordial fused model differentiated into several significantly-independent mechanisms. In animal minds emotion-motivation, conceptual understanding, and behavior-voicing have been undifferentiated unity, their differentiation is a hallmark of human evolution. Was this a single step, or could evolutionary anthropology document continuous process of differentiation or several steps, when different parts of the model differentiated from the primordial whole?

This chapter has *solved* several principled mathematical problems, which involved combinatorial complexity, when using previously considered mechanisms inspired by logical intuition. Still much remains unknown, much research and development belongs to future; some of these have been and still is going to be discussed throughout the chapter. Here we summarize some of the experimental evidence and remaining challenges for the theory discussed so far. We address three fundamental aspects of the theory: **KI, DL, the dual model**, and discuss how they relate to concepts popular in the field: recursion, and arbitrariness of vocalization. Let us address each of this in turn.

Initial experimental evidence supports **KI** (Levine & Perlovsky, 2008). Let's remind that NMF mathematically models KI as maximization of similarity between internal models and incoming sensor signals. Its psychological-cognitive interpretation is the mechanism of matching bottom-up and top-down neural signals. Satisfaction or dissatisfaction of KI is felt as aesthetic emotions, the foundations of all our higher mental abilities. Experimental proofs that such emotions actually exists, therefore is of paramount significance for much of psychology and cognitive science. Recent experiments demonstrated that these aesthetic emotions related to knowledge actually exist (Perlovsky, Bonniot-Cabanac, & Cabanac, 2010). Our theory resulted in predicting the fundamental nature of aesthetic emotions at higher levels of the mind (including emotions of the beautiful) and their role in abstract processes of higher cognition. Experimental studies of aesthetic emotions at higher cognitive levels is a challenge for the future. Alternative exploration of these mechanisms can proceed through mathematical simulations (Artificial Life).

DL is a computational technique maximizing KI without combinatorial complexity. Its mathematical foundation is a process "from vague to crisp." We discussed a simple experiment with closed and opened eyes demonstrating that in visual perception initial mental representations are vague, that perception proceeds form vague to crisp and therefore DL models the brain perception processes. Detailed neuro-imaging experiments (Bar et al, 2006) demonstrated the process "from vague to crisp"; and further confirmed that DL models the brain mechanisms of perception. This experiment further demonstrated that during visual perception this process takes about 160 ms; that this perception process occurs unconsciously; and that only the final state of this process is available to consciousness. This final state, according to the DL prediction, corresponds to approximately logical crisp perception. It corresponds to the everyday conscious experience.

A fundamental consequence of this is the unconscious logical bias: lay people and most scientists think of the mind as an essentially logical device; decision making process might be correct or incorrect, but it is considered as essentially logical and conscious. Most publications and discussions in psychology and cognitive science proceed under this bias (unconsciously) and this lead to many difficulties, some of which we address in this book. Similarly, most engineering algorithms are designed under the same bias of solving problems by conscious logical steps.

Future research will demonstrate the DL neural mechanisms for a large variety of objects, for perception in contexts and perception of situations applied to real sensor data. Neural experiments will verify the prediction of DL for the meaning of *vagueness* for situations. When perceiving situations, initial situation representations are vague in that objects are vague, and in that objects are *initially associated* with more than one situation. Experiments would probe more details of top-down and bottom-up signal interaction. The next step would extend these experiments to higher levels of cognition.

A support for the **dual model** comes from Arbib (2005). As we have mentioned, this publication have suggested a "language prewired brain" hypothesis: the mirror neuron system neurally connects motor and cognitive areas of the brain to the brain language area, according to the dual model. Another support for connections between language and cognitive circuits comes from (Franklin et al, 2008), where it was demonstrated that learning a word for a color rewires perception of this color from right to left hemisphere (where language mechanisms are located). Future experiments should demonstrate that language models guide learning of abstract cognitive models. The dual model predicts that mental representations of abstract *cognitive* contents (models) in children *remain vague longer* than representations of concrete contents directly available to perception (objects), *and longer than language* representations of abstract contents. This process, of learning crisp cognitive representations according to language, continues throughout life. The dual model predicts that *cognition* of complex abstract representations remains vaguer than the related *language* content throughout life (except specific areas of personal expertise). This prediction is easy to test by monitoring brain areas involved in cognition. Consider a subject reading a text, which switches from everyday objects to abstract contents. Our prediction is that everyday objects will stronger excite visual imaginations relative to language areas, than abstract ideas will excite cognitive areas relative to language areas.

This chapter *challenges* the idea that **recursion** is the main fundamental mechanism setting human language apart from animal abilities. This chapter proposes instead that recursion is accomplished by the hierarchy: every next level in the hierarchy accomplishes recursion of lower level models. The dual model is a simple neural mechanism, the fundamental mechanism of the mind, enabling the hierarchy, recursion, and connection of language and cognition.

The paper also *challenges* the idea of **arbitrariness of vocalization**. The arguments were first discussed in details in Plato dialogue Cratylus. The two opposite points of view were discussed there; first, that sounds of words relate to their meanings arbitrary, and second, that there are naturally predetermined relations between sounds and meanings. Recently, given the fact that in thousands of existing languages similar meanings are expressed by differently sound words, many mathematicians, philosophers, and linguists have suggested that the sound of a word is related to its meaning arbitrarily. The dual model and evolution of languages suggest a different view: a significant degree of arbitrariness in current languages is a distal result of millennia of language evolution in the presence of the dual model. Instead of assuming arbitrariness as fundamental, future research should concentrate on its emergence from the primordial fusion of sound, emotion, and meaning. We suggest that the animal fusion, the genetically fixed relations between vocalizations and meanings-emotions, has been replaced in language evolution by *habitual relations*. In some languages these habitual connections persist over centuries and millennia, in other languages they change over a lifetime. In the following sections we discuss the language mechanisms affecting the speed of these changes, and how this speed affects language and cultural evolution.

Emergence of the hierarchy is an unsolved problem. We suggest that the hierarchy is not inborn: it seems unreasonable to imagine that the number of hierarchical levels in languages and cultures is fixed genetically. We suggest further that the hierarchy evolves in cultural evolution under the drive of KI. The dual model coordinates the hierarchy of cognition and language. The ontological emergence of the hierarchy (in every individual life) is driven by the surrounding language and dual model; individual differences in creativity are due to KI. As we discuss below, fundamental differences among languages are due to different language emotionalities. In the following sections we address mathematical modeling of the evolution of hierarchy. The above proposals should be modeled mathematically and demonstrated experimentally.

We have not addressed lower hierarchical levels, below words and objects. This is another area for future research. We suggest that some aspects of these mechanisms, especially muscles of the vocal tract, can be modeled using parametric models similar to those in sections 3.1 through 3.6, with model parameterization being inborn, and other aspects can be modeled by using general "situation" models of section 3.7. Situation models suggest that words are built from phonemes similar to situations built from objects, with additional temporal relations. Similarly, objects are built from perceptual features with addition of spatial relations. These mathematical models should be developed, dynamic logic should be integrated with ongoing development in this area (see Guenther 2006).

Mathematical simulations of the proposed mechanisms should be extended to the engineering developments of Internet search engines with elements of language understanding. The next step would be developing interactive environments, where computers will interact among themselves and with people, gradually evolving human language and cognitive abilities—some aspects of this development are addressed later.

4.5 Symbols: Grounded, Perceptual, and Amodal

Symbols are not entities, but processes. We consider mathematical models of amodal, grounded, and perceptual symbols. Amodal symbols we relate to classical logic and static signs, whereas grounded perceptual symbols we relate to dynamic processes in the brain.

The idea of a symbol, a mental entity standing for another entity, is a most fundamental one making the difference between the human and pre-human minds. Yet, "Symbol is the most misused word in our culture" (Terrence Deacon, 1998). We use this word in trivial cases referring to traffic signs, and in the most profound cases of cultural and religious symbols. Valid mathematical models of symbols are fundamental for understanding workings of the brain-mind and for developing efficient computational intelligence procedures. In this section we discuss scientific understanding of what symbols are and develop appropriate mathematical models.

4.5.1 A Bit of History

Charles Peirce (1897, 1903) considered symbols to be particular type of signs. Ferdinand De Saussure (1916) emphasized that the *sign* receives meaning due to arbitrary conventions, but *symbol* implies motivation, and therefore is emotional and purposive. These arguments were forgotten during the Cognitive Revolution in the middle of the last century, cognitive scientists were inspired by new forms of representation based on developments in logic, computer science, linguistics, and statistics. They adopted abstract representations, such as feature lists, semantic networks, and frames (Barsalou & Hale 1993). These abstract representations are not related to the brain-mind mechanisms of perception, not related to any sensory *mode,* and later they received a name of *amodal* representations. Higher cognitive abilities, including the fundamental among them, symbolic ability, was assumed separate from lower level perceptions. Long ago these methods have become outdated in computational intelligence. However, computational lessons have been slowly learned in psychology and cognitive science. Computational ideas inspiring psychological thinking about the brain mechanisms of *symbols*, remain outdated by decades.

Little empirical evidence supports amodal symbolic mechanisms (Barsalou 1999). It seems, amodal symbols were adopted largely because they promised to provide "elegant and powerful formalisms for representing knowledge, because they captured important intuitions about the symbolic character of cognition, and because they could be implemented in artificial intelligence." As we have discuss in chapter 1, these promises were unfulfilled, they faced fundamental mathematical difficulties. And as we have discussed in this chapter, amodal symbols described by classical logic are final states of the dynamic logic processes. Their long-standing influence on scientists is due to properties of consciousness. Whereas the dynamic logic processes are inaccessible to consciousness, their final states, amodal symbols are conscious and make up the entire content of consciousness.

Grounded cognition seeks to ground cognitive functions in perception processes. It includes cognitive linguistics theories; theories of situated action; theories grounding cognition, memories, actions, language, and symbols; and cognitive simulation theories, in particular perceptual symbol system (PSS, Barsalou 1999), on which we concentrate in this section. PSS grounds cognition in perception. "Grounded cognition… rejects the standard view that amodal symbols represent knowledge in semantic memory." This publication emphasized the roles of simulation in cognition. "Simulation is the reenactment of perceptual, motor, and introspective states acquired during experience with the world, body, and mind… when knowledge is needed to represent a category (e.g., chair), multimodal representations captured during experiences … are reactivated to simulate how the brain represented perception, action, and introspection associated with it."

4.5.2 DL of PSS: Perceptual Cognition, Simulators, Symbols, and Signs

Simulation is an essential computational mechanism in the brain. The best known case of these simulation mechanisms is mental imagery (e.g., Kosslyn 1980; 1994); other forms of grounded cognition include situated actions, social and environmental interaction (e.g., Barsalou 2003a; Barsalou et al. 2007; Yeh & Barsalou 2006). We would emphasize that imagery is a subset of simulation; it includes various sensory-motor and emotional signals, and its dynamic aspect in PSS is usually not available to consciousness. According to PSS cognition supports action. Simulation is a central mechanism of PSS, yet rarely, if ever, they recreate full experiences. Using the mechanism of simulators, which approximately correspond to concepts and types in amodal theories, PSS implements the standard symbolic functions of type-token binding, inference, productivity, recursion, and propositions. Using these mechanisms PSS retains the symbolic functionality. "Thus, PSS is a synthetic approach that integrates traditional theories with grounded theories." (Barsalou 1999; 2005; 2007). Here we develop a mathematical model of the PSS theory using dynamic logic. We argue that the central processes in PSS, simulators, are modeled by DL. By connecting vague and unconscious representations to crisp and conscious ones, DL, similarly to the PSS simulators, connects grounded cognition to amodal representations. Similarly to Barsalou we suggest simulators modeled by DL are the symbol processes. The DL processes model simulators, the brain processes of interacting top-down and bottom-up signals. A simulator creates top-down signals from a stored mental representation; in this process it recreates salient aspects of the past bottom-up signals, which were used to generate this representation. These simulated signals are matched to the current ones. Multiple simulators compete for the best match due to the DL competition mechanism described in chapter 2. In this process mental representations are modified for the best match, new representations are created as needed. Thus PSS simulators, modeled by the DL symbol processes create new knowledge and connect mental representations to other entities—mental or in the surrounding world.

Simulators, dynamic symbol processes, when successfully matched to bottom up signals produced by the world or mental events, results in the new mental representations, approximately amodal pointers to these other entities. Simulators, modeled by the DL processes, are driven by KI, they are motivated to increase knowledge, and closely related to aesthetic emotions; in other words these processes are symbols. Their final states, approximately amodal, logical, unmotivated pointers to these events, should be appropriately called not symbols, but signs, like marks on a paper pointing to real events.

Section 4.3 extended symbol processes to unified operation of cognition and language. The same general DL technique of section 3.7. can be applied to higher, abstract hierarchical levels. Abstract cognitive representations are not grounded in direct perceptions. They require grounding in language, which is accomplished through the mechanism of the dual model. As discussed, language representations are learned ready-made from a surrounding language; therefore since young age they become crisp and conscious, stationary amodal signs. This crispness of language representations hide vagueness of cognitive representations from consciousness. Without special cognitive effort, simulators may not function adequately, cognitive representations remain vague for life. Thus, higher up in the hierarchy, the role of simulator-processes and dynamic symbols tend to diminish; and static, approximately logical, amodal signs become more important, and more accessible to consciousness. This adds to confusion between dynamic symbol processes and static sign states.

Here we briefly recollect previous discussions. Sections 2.1 through 2.6 have illustrated DL for recognition of simple objects in noise; these are complex engineering problems, unsolvable by prior state-of-the-art algorithms, still too simple to be directly relevant for PSS. Section 3.7 considered a problem of situation learning, assuming that object recognition has been solved. We recollect the principled difficulty: every situation includes many objects that are not essential to recognition of this specific situation; in fact there are many more "irrelevant" or "clutter" objects than relevant ones. Combinations of even a limited number of objects exceed what is possible to learn in a lifetime as meaningful situations and contexts from random sets of irrelevant objects. The DL technique in section 3.7 solves this difficulty. Previous sections in this chapter discuss that this solution is mathematically general and appropriate for matching top-down and bottom-up signals at every hierarchical level. In other words, it models symbol processes and creation of mental representations at each hierarchical level, with the remaining difficulty of grounding. Similarly this technique leads to learning of language from surrounding language. Combining DL and the dual model, perceptual symbol simulators lead to learning of abstract cognitive representations guided by (grounded in) language. This process however is not as straightforward as at lower levels. At higher abstract levels of cognition, crisp language representations hide vagueness of cognition from consciousness. Simulators may not function autonomously, and abstract cognitive representations may remain vague throughout the lifetime. At higher abstract level most people, most of the time think using language, rather than cognitive representations grounded in experience. We return to this difficulty of thinking in the next chapter 5 and connect it to a fundamental irrationality of human thinking, the subject of the 2002 Nobel Prize.

4.5.3 Other PSS Operations: Concepts, Productivity, Grounding, and Binding

Here we continue discussing relationships between mathematical DL procedures and fundamental ideas of PSS and cognitive science. PSS grounds perception, cognition, and high-level symbol operation in modal symbols, which are ultimately grounded in the corresponding brain systems. Sections 4.5.2 and 3.7 provide an initial "first step" toward developing formal mathematical description suitable for PSS. We have considered one subsystem of PSS, a mechanism of learning, formation, and recognition of situations from objects making up situations. The mind's representations of situations are signs-concepts of a higher level of abstractness than signs-objects making them up. The mechanism of matching bottom-up and top-down signals involves PSS simulators, modeled by DL. We have also discussed that all abstract concepts at all hierarchical levels can be modeled using this technique without combinatorial complexity. The proposed mathematical formalism can be advanced straightforwardly to "higher" levels of more and more abstract concepts. Similarly, the proposed mathematical formalism can be applied at a lower level of recognizing objects as constructed from their parts. Mathematical techniques of sections 2 and 3 can be combined to implement this PSS object recognition idea as described in (Barsalou 1999): objects are constructed from sensor-defined features, similar to how situations are constructed from objects. According to the described theory, DL processes "from vague-to-crisp" model PSS simulators or symbol processes; in this section we use these interchangeably.

First we address *concepts* and their development in the brain. According to (Barsalou 2007),

"The central innovation of PSS theory is its ability to implement concepts and their interpretative functions using image content as basic building blocks."

This aspect of PSS theory is implemented in DL in a most straightforward way. Concept-situations in DL are collections of objects (representations and simulators at lower levels, which are neurally connected to neural fields of object-images). While objects are perceptual entities-symbols in the brain, concept-situations are collections of perceptual symbols. In this way situations are perceptual symbols of a higher order complexity than object-symbols, they are grounded in perceptual object-symbols (images), and in addition, their learning is grounded in perception of images of situations.

Barsalou (2008) emphasized that concepts in the brain are sets of correlated features that are multimodal and distributed (perceived by various sensor and motor modalities in various parts of the brain). Neural realization of these processes is implemented in the brain by a population of conjunctive neurons (Simmons & Barsalou 2003).

Concepts refer to both, approximately amodal, logical-like, conscious representations; and concepts refer to the mechanism of recognizing the corresponding events. Also, concepts can be used in imagination, when constructing plans. Thus *concepts* sometimes refer to the DL simulator-symbol processes. The ability of concept-mechanism to create imaginations is referred to as *productivity*. The described theory mathematically models productivity of the mind concept-simulator system as DL processes. Other widely used notions in cognitive science are *types* and *tokens*. Types denote a concept or class of objects or events. They are modeled as vague representations and the DL symbol processes. Tokens denote individual objects or events of the type and are modeled as final states of these processes, approximately amodal, logical-like, conscious representations. In the process of learning DL symbol simulators "interpret individuals as tokens of the type" (Barsalou 2008).

Perceptions of imagined situations in the mind, as mentioned, are the essence of imagination. Models of situations (probabilities of various objects belonging to a situation, and objects attributes, such as their locations) can depend on time, in this way they are parts of simulators accomplishing cognition of situations evolving in time. If "situations" and "time" are invoked "with closed eyes" and pertain to the mind's imaginations, the simulators implement imagination-thinking process, or planning.

Usually we perceive-understand a surrounding situation, while at the same time thinking and planning future actions and imagine consequences. This corresponds to running multiple simulators in parallel. Some simulators support perception-cognition of the surrounding situations as well as ongoing actions, they are mathematically modeled by DL processes that converged to matching internal representations (types) to specific subsets in external sensor signals (tokens). Other simulators simulate imagined situations and actions related to perceptions, cognitions, and actions, produce plans, etc.

Integrating multiple pieces, objects, events into higher level events is referred to as *binding*. This important ability of both cognition and language was extensively discussed in literature along with possible mechanisms; previously suggested mechanisms, as discussed, have led to combinatorial complexity and thus were not computable. Modeling situations in PSS using DL is a general solution of the binding problem.

DL provides a general approach to the binding problem (-> move to the end of chapter: binding in PSS: Edelman & Breen 1999; DL also mathematically models the "corkboard" approach described in Edelman & Intrator 2001).

Described here DL modeling of PSS models mathematically what Barsalou (2003b) called dynamic interpretation of PSS (DIPSS). DIPSS is fundamental to modeling abstraction processes in PSS. Three central properties of these abstractions are type–token interpretation; structured representation; and dynamic realization. Traditional theories of representation based on logic model interpretation and structure well, but are not sufficiently dynamic. Conversely, connectionist theories are dynamic but are inadequate at modeling structure. PSS and the DL mathematical process address all three properties. In type–token relations "propositions are abstractions for properties, objects, events, relations and so forth. After a concept has been abstracted from experience, its summary representation supports the later interpretation of experience." Correspondingly in the developed mathematical approach, DL models a situation as a loose collection of objects. Its summary representation (concept, the initial vague model) evolves-simulates representation of a concrete situation in the process of perception of this concrete situation. A loose collection of property and relation simulators according to DL represent abstractions. The DL process involves structure (the initial model) and dynamics (the DL process). Using DL-PSS mathematical model for symbolic predication and conceptual combinations can be developed from the above description.

4.5.4 Perceptual Symbols vs. Amodal Signs

Relations between amodal logical signs and grounded symbol processes is a topic of utmost significance. Due to a long history and loaded meanings of *signs* and *symbols* it could create misunderstandings despite much of the above discussions. Therefore this section is specifically addresses topic. Since any mathematical notation may look like an amodal symbol, we first discuss the roles of amodal vs. perceptual systems in DL. This would require clarification of the word symbol. We touch on related philosophical and semiotic discussions and relate them to mathematics of DL and to PSS. For the sake of brevity we limit discussions to general interests, emphasizing connections between signs and symbols, DL, perceptual and amodal systems. We summarize here related discussions scattered throughout the chapter. (-> to the end: extended discussions of symbols can be found in Perlovsky 2006b;d).

"Symbol is the most misused word in our culture" (Deacon, 1998). Why the word "symbol" is used in such a different way: to denote trivial objects, like traffic signs or mathematical notations, and also to denote objects affecting entire cultures over millennia, like Magen David, Swastika, Cross, or Crescent? Let us compare in this regard opinions of two founders of contemporary semiotics, Charles Peirce (Peirce 1897; 1903) and Ferdinand De Saussure (1916). Peirce classified signs into symbols, indexes, and icons. Icons have meanings due to resemblance to the signified (objects, situations, etc.), indexes have meanings by direct connection to the signified, and symbols have meaning due to arbitrary conventional agreements. Saussure used different terminology, he emphasized that signs receive meanings due to arbitrary conventions, whereas symbol implies motivation. It was important for him that motivation contradicted arbitrariness. Peirce concentrated on the process of sign interpretation, which he conceived as a triadic relationship of sign, object, and interpretant. Interpretant is similar to what we call today a representation of the object in the mind. However, this emphasis on interpretation was lost in the following generation of scientists. This process of "interpretation" seems close to DL processes and PSS simulators. We therefore follow Saussurean designation of symbol as a motivated process.

Motivationally-loaded interpretation of symbols was also proposed by Jung (1921). He considered symbols as processes bringing unconscious contents to consciousness. Similar are roles of PSS simulators and DL processes.

In the development of scientific understanding of symbols and semiotics, the two functions, understanding language and understanding world, have often been perceived as identical. This tendency was strengthened by considering logic to be the mechanism of both, language and cognition. According to Russell (1919), language is equivalent to axiomatic logic, "[a word-name] merely to indicate what we are speaking about; [it] is no part of the fact asserted… it is merely part of the symbolism by which we express our thought". Hilbert (1928) was sure that his logical theory also describes mechanisms of the mind, "The fundamental idea of my proof theory is none other than to describe the activity of our understanding, to make a protocol of the rules according to which our thinking actually proceeds." Similarly, logical positivism centered on "the elimination of metaphysics through the logical analysis of language" – according to Carnap (1959) logic was sufficient for the analysis of language. As discussed in section 2.2, this belief in logic is related to functioning of human mind, which is conscious about the final states of DL processes and PSS simulators, which are perceived by our minds as approximately logical amodal symbols. Therefore we identify amodal symbols with these final static logical states, signs.

DL and PSS explain how the mind constructs symbols, which have psychological values and are not reducible to arbitrary logical amodal signs, yet are intimately related to them. In section 3.7 we have considered objects as learned and fixed. This way of modeling objects indeed is amenable to interpreting them as amodal symbols. Yet, we have to remember that these are but final states of previous simulator processes, perceptual symbols. Every perceptual symbol-simulator has a finite dynamic life, and then it becomes a static symbol-sign. It could be stored in memory, or participate in initiating new dynamical perceptual symbols-simulators. This infinite ongoing dynamics of the mind-brain ties together static signs and dynamic symbols. It grounds symbol processes in perceptual signals that originate them; in turn, when symbol-processes rich their finite static states-signs, these become perceptually grounded in symbols that created them. We could become consciously aware of static sign-states, express them in language and operate with them logically. Then, outside of the mind-brain dynamics, they could be transformed into amodal logical signs, like marks on a paper. Dynamic processes – symbols-simulators are usually not available to consciousness. These PSS processes involving static and dynamic states are mathematically modeled by DL in section 3.7.

To summarize, DL does not model just amodal symbols, which are governed by classical logic; this would lead to combinatorial complexity. DL operates on different type of PSS representations, which are vague combinations of lower-level representations. These lower-level representations could include memory states as well as vague dynamic states from concurrently running simulators – DL processes of the on-going perception-cognition. To the extent that the mind-brain is not a strict hierarchy, the same-level and higher-level representations could be involved along with lower levels. Thus DL models processes-simulators, which operate on PSS representations. These representations are vague and DL processes are assembling and concretizing these representations. As described by Barsalou, bits and pieces from which these representations are assembled, could include mental imagery as well as other components, including multiple sensor, motory, and emotional modalities; these bits and pieces are mostly inaccessible to consciousness during the process dynamics. DL also explains how logic and ability to operate amodal symbols originate from illogical operations of PSS.

The described DL formalization of PSS, suggests using a word *signs* for amodal static logical constructs outside of the mind, including mathematical notations; and to reserve *symbols* for perceptually grounded motivational cognitive processes in the mind-brain. Memory states, to the extent they are understood as static entities, are modeled by signs in this terminology. Logical statements and mathematical signs are perceived and cognized due to PSS simulator processes; after events are understood they become signs. Perceptual symbols, through simulator processes, tie together static and dynamic states in the mind. Dynamic states are mostly outside of consciousness, while static states might be available to consciousness.

4.5.5 Experimental Evidence and Future Research

Future research will address the DL mathematical theory of PSS throughout the mind hierarchy; from features and objects "below situations" in the hierarchy to abstract models and simulators at higher levels "above situations." Modeling across the mind modalities will be addressed including diverse modalities, symbolic functions, conceptual combinations, predication. Modeling features and objects would have to account for suggestions that perception of features are partly inborn (Barsalou 1999); this development therefore might require new experimental data about which feature aspects are inborn (Edelman & Newell, 1998). The developed DL formalization of PSS corresponds to observations in (Wu & Barsalou 2009) and it will be used for generating detailed experimentally verifiable predictions. The DL formulation developed in section 3.7 has 2 hierarchical levels, objects and situations, and it demonstrates bindings within these two levels. In future hierarchical extension of the DL binding will be related to hierarchy as a general mathematical principle (Edelman 2003). Similarly, the recursive property of cognition and language will be modeled as a mathematical hierarchy.

Experimental research (Bar et al. 2006; Bar 2007) will address specific properties of higher level simulators predicted here. Among these is a prediction that early predictive stages of situation simulations are vague. Whereas vague predictions of objects resemble low-spatial frequency of object imagery (Bar et al. 2006), "the representation of gist information on higher levels of analysis is yet to be defined" (Bar 2007). According to DL, vague mental models of situations and contexts should contain many objects with low probabilities; most of these objects are not relevant to the situation. Since situation recognition and object recognition are going on in parallel, the object mental models are vague. The hierarchical DL model is applicable to higher levels ("above" object-situations), this predicts the nature of information of higher level gists, and it will be experimentally verified.

The DL model can be expanded to address another topic discussed in (Bar 2007), "how the brain integrates and holds simultaneously information from multiple points in time." Two different mechanisms should be explored: first, explicit incorporation of time into models (so that model parameters and probabilities depend on time), and second, categorized temporal relations, such as "before," "after" are to be included, similar to any other relations into models. A joint mathematical-experimental approach will be fruitful in this area.

Future research will address grounded symbols in view of the previous section model of interaction between language and cognition. Since language models are acquired "ready-made" from surrounding language, rather than from life experience, language is closer aligned with amodal symbols than with perceptual symbols. Kids at 5 years of age can talk about much of cultural content of the surrounding language, including highly abstract contents; yet, clearly kids do not have necessary experience to understand highly abstract concepts as perceptual symbols, and to relate them to the world. Future research will address origin of amodal symbols in language. The DL model of language-cognition interaction proposed in the previous section suggests that higher abstract concepts could be stronger grounded in language than in perception; not only kids, but also adults may operate with abstract concepts as with amodal symbols, and therefore have limited understanding grounded in experience of how abstract concepts relate to the world. Future experimental research should address this hypothesis.

4.6 Future Man-Machine Systems

Future man-machine systems will be ubiquitous and omnipresent, from pilot cockpits to living rooms, to mobile telecom and entertainment devices. They will have cognitive computational abilities described in this book, and they will interact with us, their users, with increasingly human-like modes. They would learn from us language and use it for communication, they would learn our habits and understand our needs and desires.

4.6.1 Cooperative and Interactive Systems

In current man-machine systems, bottlenecks and weak links are the interfaces. Our thinking and machine computational abilities are much faster than input-output processes. The crucial speed up would be possible when machines will understand human language. Current language understanding devices are unreliable, non-robust, non-adaptive to individual users. The main weakness of the current devices is related to the engineering principles they are based on. These principles have been successful for building cars and airplanes but are inadequate for the next level of cognitive devices that would have to approach human level functioning.

Future interactive systems will approach human level abilities for language and cognition by using techniques described in this chapter, particularly in sections 4.3 and 4.4. They would not require extensive programming, they will learn language and cognition by interacting with humans and among themselves. Instead of becoming obsolete with time they will become smarter by accumulating knowledge.

4.6.2 Semantic Web

Future semantic web will be one of these cooperative interactive systems. Contemporary web mostly stores information. Two features make it much more useful than data bases of the past. First, everyone can create a webpage and post their data and information of potential interest to others. Second, search engines make this information accessible to users. Limitations of current search engines are the main impediment to usefulness of the web. Google and Yahoo do not understand language. Their interpretation of queries does not match human needs. Often they respond with thousands of pages, or cannot match a request to a single page. Language learning technique of section 4.2 will significantly enhance abilities of search engines. To improve use of imagery and video on the web, search engines should be added abilities of combined language and cognition described in section 4.3.

These search engines with abilities for language and cognition will become intelligent agents with abilities approaching human mind. The next step will be personal agents that will gradually replace personal web pages. They will have access to personal computers, and they will interact with each other on the web, thus expanding the human noosphere to a next level.

Personal intelligent agents will acquire emotional abilities, which we discuss in the next section.

4.7 Emotional Intelligence and Love from the First Sight

Emotions and their role in cognition. Emotions are related to instinct satisfaction or dissatisfaction. Throughout this book we consider in details one instinct KI, and we consider emotions related to this instinct, aesthetic emotions related to knowledge. There are few basic instincts and emotions, but there are zillions of aesthetic emotions. Interaction of conceptual and emotional intelligence leads to love from the first sight.

4.7.1 Emotions

Psychologists and neuro-psychologists identify several psychic processes and neural mechanisms referred to as emotions. Cabanac (2002) emphasized that there is no consensus on a definition of emotion and define these as motivational states with hedonic contents (pleasure-displeasure). Russell & Barrett (1999) and Russell

(2003) called these undifferentiated motivational states "core affect." Undifferentiated emotions are perceived along two axes: strong-weak (arousal) and good-bad (valence). Juslin and Västfjäll (2008) emphasized a number of neural mechanisms involved with emotions and different meanings implied for the word 'emotion'.

In this book we follow (Grossberg & Levine 1987) and consider emotions as neural signals and corresponding subjective feelings, which indicate to the brain-mind satisfaction or dissatisfaction of instinctual needs. Primitive animal organisms might have just one undifferentiated emotion-feeling of good-bad, characterized only by arousal and valence (good-bad). Humans can differentiate many emotions. Emotions are differentiated according to the instinctual needs they indicate: we feel a need for food (hunger) differently from a need for water (thirst), and differently from a need for safety (scare). Emotions could be differentiated further: needs for different types of food could be felt differently. We refer to emotional differentiation as quality of emotions.

Differentiated emotions are accessible to consciousness in more details than undifferentiated ones. Similarly to concepts that are accessible to consciousness in details corresponding to how well they are differentiated, quality of emotions corresponds to their differentiation in consciousness. Psychologists usually discuss few "basic" emotions, corresponding to basic instincts. We have words for these emotions. These words are not necessarily same in different languages. But even more fascinating are zillions of emotions related to KI, and later in this chapter we discuss these emotions, their functions and mechanisms in details. Here we would just emphasize that we hear these emotions in language sounds, in music, in songs. Even so we do not have special words for most of these emotions, still we use these emotions for making decisions and for behavior. Spinoza (2005/1677) was the first thinker who noted that emotions differ depending on which object they refer to.

Let us repeat, in this book we consider in detail only one instinct, the instinct for knowledge (KI). It is modeled mathematically by maximization of similarity between bottom-up and top-down signals. Emotions corresponding to KI are modeled by similarity changes; they are called aesthetic emotions. Our mathematical model of KI in previous chapters can only model a single aesthetic emotion, satisfaction or dissatisfaction of KI, felt emotionally as harmony or disharmony (between mental models and the world); this is modeled by changes in the similarity measure. This is clearly inadequate to model a huge variety of aesthetic emotions. A psychologically valid model of KI is differentiated, it is not a single measure of similarity between all our mental representations and all our experience. Every piece of knowledge may correspond or contradict to basic instinctual needs and involves aesthetic emotions related to *understanding* (in addition to entirely different, basic emotions that involve satisfaction or dissatisfaction of the basic instincts). Even more differentiation is involved in knowledge, understanding of virtually every combination of several pieces of knowledge could be felt emotionally and involve a different aesthetic emotion.

In section 4.2 we discussed an aesthetic emotion at the very top of the mind hierarchy. At this top level knowledge refers to the meaning of the entire life experience. The differentiation status of this knowledge and related emotions is complex. Previously we emphasized that this knowledge is vague, undifferentiated, unconscious, and satisfaction of KI at the highest level is felt as undifferentiated emotion of the beautiful. Now we would like to emphasize that this is just one aspect of the emotion of the beautiful. Even so the beautiful is not crisply differentiated, the more rich and diverse is one's system of knowledge and emotions, the richer are shades of the understanding of the life meaning and of the emotion of beautiful one can perceive.

At the bottom of the mind hierarchy, KI operates autonomously, and emotions related to its satisfaction or dissatisfaction, say, during perception of an everyday object, are below the threshold of consciousness. At higher levels, need for knowledge and related aesthetic emotions are conscious. These emotions are differentiated. We feel differently about satisfying our need to know if we can rely on our acquaintance's minor causal comment, how to select vine at a party, how to make vacation plans, how to invest money, how to choose schools for kids, or when solving a problem for our Ph.D. thesis after two years of effort. As mentioned, our previous definition of KI as maximizing a single measure of similarity between all our mental representations and all our experiences is inadequate, too simplistic. KI is a differentiated ability and developing a mathematical model for this differentiated KI is a problem for future research. Knowledge is not a single undifferentiated measure. It is a huge number of differentiated measures of similarities between various bottom-up and top-down signals. Because of the hierarchy of the mind, it involves different-size "chunks" of knowledge: at the object level, KI strives to understand objects, at a higher situational level, KI strives to understand groups of objects, etc. Every piece of knowledge involves its own aesthetic emotion, and every combination of different pieces of knowledge involves an emotion evaluating understanding of relations between these pieces of knowledge. People differ in their consciousness and in abilities to differentiate these emotions.

4.7.2 Intelligence

Even so this book is to a significant extent about intelligence, most likely, intelligence cannot be fully defined and discussions in this section outline multiple directions for future research. In this section we briefly discuss intelligence from two sides: starting from the lowest intelligence as it evolved from pre-life, and starting from the highest human intelligence. This cursory discussion is aimed at outlining questions and directions along which the answers should be thought. The "definition" of what is intelligence will come after complex multi-agent models of the mind, societies, cultures will be developed and studied in great details. Here we outlined a path to this development.

In previous chapters we defined intelligence as a similarity measure between mental representations and surroundings. We considered similarity measures related to likelihood (of a particular observation, given certain mental models) and to mutual information (in mental models about the world). We did not address a foundational mathematical question that each of these quantities require, identifying *states* of two systems: the mind and the world. Only when elementary states are defined, one can define probabilities and information.

Models, or mental representations answer this first question, they define "elementary states" within the mind. At the bottom of the mind hierarchy, sensory-motor mechanisms define the elementary states sensed in the world: what can be perceived by sensors and operated on behaviorally. At every higher level of the mind hierarchy, mental representations define states at this level. In the animal world mental representations do not go higher than objects and situations—what can be directly perceived.

The next question, where these representations have come from? Some representations are "better" than others, and the next question: better for what purpose? It seems at this point we have to consider the entire evolution. Dawkins thinks that gene survival and propagation is the ultimate answer, and genes are "the representations". We disagree. Evolution and intelligence are different mechanisms with different purposes. Evolution favors most simple animals, or even viruses—there are much more genes inside viruses, than inside human beings. Yet, Dawkins cannot explain the "elephant in the room": why individual intelligence (of chickens, dogs, or humans) increases in evolution.

Our explanation is that at some point in evolution individual organisms appeared. We define it as divergence between "gene interest" to replicate and propagate and "individual interest" to survive. Much is written about individual interests being defined to a significant extent by genes, yet, even the very existence of these discussions indicates differences. Later in evolution, beginning possibly with amniotes (reptiles, birds, etc.), adaptive mental representations evolved. Adaptation required the knowledge instinct, KI, a drive to better understand the environment and self and the interaction between the two. Initially it was for the purpose of survival, but survival of the individual or family, not genes per se. In humans, KI goes beyond survival of the humankind—which is proven by an ability of the humankind to destroy itself. Does it make sense to talk about KI at a purely genetic level? – Possibly genetic evolution since its earlier stages where driven by developing molecular structures, which could identify complementary structure in the surrounding world of biological molecules. We do not address dynamic logic of molecular evolution in this book. We would just mention that evolution through mutations is as unlikely as evolution of mind through logical searches. Today geneticists assume that genetics mechanisms are random searches. It may not be correct, but this is the current state of knowledge. Do lobsters have the knowledge instinct? Possibly not. By the time individually-adaptive internal representations appeared, KI had to be present to drive the adaptation. Adaptation toward what? Goals of adaptation include survival, but do not stop at that. In this book we do not consider connecting knowledge to survival and do not consider mechanisms KI mechanisms of genetic evolution.

Another approach to intelligence starts "from the top" and attempts to understand various human intelligences, as various abilities to understand the surroundings and self, and to use this understanding advantageously. What are these intelligences, and what does it mean to use them advantageously? There is the whole branch of psychology measuring various answers to these questions. We touch it briefly.

Intelligence includes abilities to understand the surrounding conceptually as pieces of knowledge, to evaluate it emotionally, to make decisions, and to control one's body and mind in processes of achieving goals of these decisions. Intelligence cannot be defined as a single measure. Psychologists discuss multiple intelligences. People have different abilities in different areas. Some are good at poetry, others at math, at money investments, at choosing friends, or life partners. Intelligence could be seen as an ability to make beneficial choices. What one considers beneficial are also choices requiring intelligence. Among the most important choice everyone has to make is to understand one's own abilities and interests. What is intelligence, how to measure it, is a separate field of study, and we are not going to consider it in all its varieties in this book. Intelligence is often associated with conceptual cognition. Standard IQ, SAT, GRE tests measure verbal and quantitative abilities. Here we would like to emphasize that identifying intelligence with conceptual abilities is inadequate. Emotional aspects of intelligence are no less important.

Discussing intelligences is closely related to understanding various psychological types and personality traits. Among important personality traits psychologists consider experimentally discovered "Big Five" factors: 1) Extraversion–Introversion, 2) Neuroticism–Stability, 3) Openness–Closedness, 4) Agreeableness–Disagreeableness, and 5) Conscientiousness–Carelessness (Tupes & Cristal 1961).

4.7.3 Emotional Intelligence

Emotional intelligence (EI) as a separately measurable ability was introduced by (Mayer 1999; Mayer, Salovey, & Caruso 2008). "Some individuals have a greater capacity than others to carry out sophisticated information processing about emotions and emotion-relevant stimuli and to use this information as a guide to thinking and behavior." They discussed four components of EI: 1) perceiving emotions accurately in oneself and others; 2) using emotions to facilitate thinking; 3) understanding emotions, emotional language, and the signals conveyed by emotions; 4) managing emotions so as to attain specific goals. To consider EI a specific form of intelligence, different from other intelligences, it is important that experiments have found it relatively uncorrelated with IQ (correlation about 0.35) and Big Five (correlation less than 0.25). Developers of EI did not concentrate on prime vs. aesthetic emotions.

We would like to emphasize that according to our previous analysis consciousness is closely related to differentiation. Better differentiated psychic functions are more conscious. Higher EI corresponds to better differentiated emotions. Understanding emotions in details and using them for achieving goals requires differentiating large number of emotions, feeling in details their connections to surroundings and to goals inside self. To achieve goals, one needs an ability to evaluate one own goals emotionally and select proper goals; to channel one's own emotions toward goals; to understand in details emotions of others, react to them with correct emotions, and communicate one's emotions precisely. A mere awareness of undifferentiated positive or negative feelings is insufficient, however strong these feelings might be; strong undifferentiated emotions are more likely to mislead than to lead toward meaningful goals.

Operations of KI involve both, conceptual and emotional aspects. Conceptual aspect, in our mathematical formulation, involves models and their parameters. Emotional-motivational aspect involves the KI and its maximization, DL. Future models of KI, to model EI, would have to develop a mathematical model of the differentiated KI, accounting for a diversity of aesthetic emotions. Here we just would mention that maximization of conditional similarities, l(n|m), might be related to differentiated KI, however, it is not an adequate model. The reason is that maximizing conditional similarities separately from each other would eliminate competition among models. This competition is essential for selecting models best describing bottom-up signals. Possibly EI related to emotionally differentiated KI can be mathematically modeled by adaptive relations-weights among sets of concepts making up a concept of a higher level.

4.7.4 Love from the First Sight, Divorce, and Other Miseries

Interactions between emotional intelligence (EI) and conceptual intelligence (CI), involve unconscious and often create poorly understood psychic states, which mislead individuals in their most fundamental life choices. To achieve happy creative life, to find the area of one's own unique abilities, to find life partners mutually enhancing each other lives, one need to understand his or her own strengths in EI and CI, to identify these strengths consciously, and consistently follow one's strengths and avoid rush judgments from positions of one's psychic weakness. Tricky underwater currents usually make this task exceedingly complex. Here we analyze these using the developed scientific knowledge and identify potential pitfalls.

Those strong in CI—usually scientists, most of the readers of this book—easily understand a large number of differentiated concepts, adequate for evaluating everyday surroundings in great details. These differentiated concepts are fully conscious and easily manipulated for achieving meaningful goals. However, emotions are "opposite" to concepts, which is well known psychologically and reflected in a large number of "folk psychology" theories and proverbs. In terms of the previous section analysis, it means that it is difficult to keep in consciousness both differentiated concepts and differentiated emotions. Scientists and other people with strong CI, usually are low in EI, their emotions are not differentiated and not crisply conscious.

And here the nature of our conscious and unconscious plays a bad trick with our psyche. For high-CI people, their many differentiated, original, well adapted concepts are psychologically easy, they come up to their minds naturally, they are conscious and therefore do not disturb our unconscious. Therefore there is a tendency to consider them as less important part of psyche, to disregard them. On the opposite, primitive undifferentiated emotions are not conscious, and in the depths of unconscious they affect primordial, less voluntary parts of psyche. They get to the guts. So many CI people tend to disregard their well adapted concepts, and instead value their primitive emotions, learned in childhood, from friends, etc. Emotions that are not adapted to their personal circumstances take over the psyche and considered the "true self." And one may select a wrong area of study and work, an inborn physicist wants to be a poet, etc. And even worse, often one makes wrong decisions in personal life.

When young, meeting an opposite, EI person, a CI person is fascinated by the opposites, by psychic features that fill personal voids. This is the psychological basis for the first love, love from the first sight. But keeping devotion to a person opposite from you, requires sustained effort. Quite often as life goes by and everyday life chores keep mounting, the same psychic qualities, instead of fascinating, turn opposite. Manipulations with diverse emotions, which are so easy for an EI person, start looking hollow, artificial, and non-genuine. They are perceived as personally manipulative. It happens exactly because for a CI these are difficult. Shallow, commonplace, poorly adapted, and non-original emotions of a CI person get him or her by the guts, therefore an easy manipulation with varying emotions are perceived as shallow. The very appropriate life partner may look like a wrong one.

The same case may look similar from the opposite site. An EI person, meeting a CI one, first is fascinated by the opposite and falls in love. Later, easy manipulation by conceptual thinking, which is so natural for a CI person, annoys an EI one. Concepts are difficult for an EI person and get her or him by the guts. The opposite ability of a life partner seems shallow, not genuine. Both come to a wrong conclusion. Divorce.

The described interaction of EI and CI explains majority of divorces. Misunderstandings of oneself, taking one's weak, unconscious abilities for the essence of self, and disregarding what is unique, conscious, adaptive, and God-given lead to uncountable miseries. It is difficult indeed to fulfill the 6th c. BCE pronouncement by the first philosopher Thales: "Know thyself."

4.8 Emotionality of Languages and Meanings

In current man-machine systems, bottlenecks and weak links are interfaces. To overcome this limitation, man-machine systems should be able to learn language and communicate with human users using language. In sections 4.2 and 4.3 we described how to overcome combinatorial complexity in computational systems learning language and to combine language with cognition. We addressed conceptual contents of languages. Now we discuss the role of emotions in language. Future systems will have to be able to learn and use language with its conceptual and emotional aspects. To understand the role of emotions in language, we have to start from pre-language animal vocalizations.

4.8.1 *Primordial Undifferentiated Synthesis of Psyche*

Animals' vocal tract muscles are controlled mostly from the ancient emotional center (Lieberman, 2000). Vocalizations are more affective than conceptual. Mithen (Mithen, 2007) summarized the state of knowledge about vocalization by apes and monkeys. Calls could be deliberate, however their emotional-behavioral meanings are not differentiated; primates cannot use vocalization separately from emotional-behavioral situations; this is one reason they cannot have language.

Emotionality of voice in primates and other animals is governed from a single ancient emotional center in the limbic system (Deacon, 1989; Lieberman, 2000; Mithen, 2007). Cognition is less differentiated than in humans. Sounds of animal cries engage the entire psyche, rather than concepts and emotions separately. An ape or bird seeing danger does not think about what to say to its fellows. A cry of danger is inseparably fused with recognition of a dangerous situation, and with a command to oneself and to the entire flock: "Fly!" An evaluation (emotion of fear), understanding (concept of danger), and behavior (cry and wing sweep) – are not differentiated. Conscious and unconscious apparently are much less separated than in humans. Recognizing danger, crying, and flying away is a fused concept-emotion-behavioral synthetic form of cognition-action. Birds and apes can not control their larynx muscles voluntarily.

This primordial synthesis of psyche makes *meaningful* every aspect of psychic life. Behavior of an animal might be not as smart as human's but it is always motivated to achieve a goal important in animal's life. An animal is incapable of meaningless behavior.

4.8.2 *Language and Differentiation of Emotion, Voicing, Cognition, and Behavior*

Origin of language required freeing vocalization from uncontrolled emotional influences. Initial undifferentiated unity of emotional, conceptual, and behavioral-(including voicing) mechanisms had to separate-differentiate into partially independent systems. Separation of voicing from emotional control was paralleled by development of a separate emotional center in cortex which controls larynx muscles, and which is partially under volitional control (Deacon, 1989; Mithen, 2007). In contemporary languages the conceptual and emotional mechanisms are significantly differentiated, as compared to animal vocalizations. The languages evolved toward conceptual contents, while their emotional contents were reduced.

Emotions, as we discussed, indicate satisfaction or dissatisfaction of instinctual needs. Reduction of emotional contents implies reduction of motivation. We return to the discussion of motivation in human language and behavior throughout the chapter. Here we emphasize that differentiation of emotions in humans opens opportunities for sophisticated motivations, but at the same time creates a possibility for unemotional, unmotivated behavior, for the loss of meanings.

4.8.3 Grammar, Language Emotionality, and Meanings

Language and voice started separating from ancient emotional centers possibly millions of years ago. Nevertheless, emotions are present in language. Most of these emotions originate in cortex and are controllable aesthetic emotions. Their role in satisfying the knowledge instinct is considered in the next section. These emotions make human behavior *motivated* full of complex *meanings*. Emotional centers in cortex are neurally connected to old emotional limbic centers, so both influences are present. Emotionality of languages is carried in language sounds, what linguists call prosody or melody of speech. This ability of human voice to affect us emotionally is most pronounced in songs. Songs and music are addressed in section 4.11.

Emotionality of everyday speech is low, unless affectivity is specifically intended. We may not notice emotionality of everyday "non-affective" speech. Nevertheless, "the right level" of emotionality is crucial for developing cognitive parts of mental models. If language parts of models were highly emotional, any discourse would immediately resort to blows and there would be no room for language development (as among primates). If language parts of models were non-emotional at all, there would be no motivational force to engage into conversations, to develop language models. The motivation for developing higher cognitive models would be reduced. Lower cognitive models, say, for object perception would be developed because they are imperative for survival and because they can be developed independently from the language, based on direct sensory perceptions, like in animals; they are motivated by primitive instinctual needs. But models of situations and higher cognition are developed based on language models, as discussed in section 4.3. This requires emotional connections between cognitive and language models. This is one aspect of meanings of aesthetic emotions, specifically human meaning, unavailable to animals. Understanding mechanisms of these emotions and meanings is essential for understanding higher level cognition, which separates us from animals.

Primordial fused language-cognition-emotional models have differentiated long ago. The involuntary connections between voice-emotion-cognition have been much reduced with emergence of language. They have been replaced with habitual connections. Sounds of all languages have changed; still some sound-emotion-meaning connections in languages remain. If the sounds of a language change slowly the connections between sounds and meanings persist and consequently the emotion-meaning connections persist. This persistence is a foundation of meanings because meanings imply motivations. If the sounds of a language change too fast, the cognitive models are severed from motivations, and meanings may disappear. If the sounds change too slowly the meanings are nailed emotionally to the old ways, and culture stagnates.

This statement is a controversial issue, and indeed, it may sound puzzling. Doesn't culture direct language changes or is the language the driving force of cultural evolution? Direct experimental evidence is limited; it will have to be addressed by future research. Theoretical considerations suggest no neural or mathematically plausible mechanism for culture directing evolution of language through generations; just the opposite, most of cultural contents are transmitted through language. Cognitive models contain cultural meanings separate from language, but transmission of cognitive models from generation to generation is mostly facilitated by language. Cultural habits and visual arts can preserve and transfer meanings, but they contain a minor part of cultural wisdom and meanings comparative to those transmitted through language. Language models are major containers of cultural knowledge shared among individual minds and collective culture.

The arguments in the previous two paragraphs suggest that an important step toward understanding cultural evolution is to identify mechanisms determining changes of the language sounds. As discussed below, changes in the language sounds are controlled by grammar. In inflectional languages, affixes, endings, and other inflectional devices are fused with sounds of word roots. Pronunciation-sounds of affixes are controlled by few rules, which persist over thousands of words. These few rules are manifest in every word. Therefore every child learns to pronounce them correctly. Positions of vocal tract and mouth muscles for pronunciation of affixes (and other inflections) are fixed throughout population and are conserved throughout generations. Correspondingly, pronunciation of whole words cannot vary too much, and language sounds change slowly. Inflections therefore play a role of "tail that wags the dog" as they anchor language sounds and preserve meanings. This, we think, is what Humboldt (1836/1967) meant by "firmness" of inflectional languages. When inflections disappear, this anchor is no more and nothing prevents the sounds of language to become fluid and change with every generation.

This has happened with English language after transition from Middle English to Modern English around the 15th c. (Lerer, 2007), most of inflections have disappeared and sound of the language started changing within each generation ("Great Vowel shift" was a part of this process), and this continues today. Among few remaining affixes are "s" for plurals and "ed" for past tense. There is [i] affix as in daddy, mommy, anty, Annie, etc., for expressing human affinity, but it is not universal, it is applicable to few words in English (e.g., in Russian and many other languages there are dozens of affixes and inflections applicable to every word). English evolved into a powerful tool of cognition unencumbered by excessive emotionality. English language spread democracy, science, and technology around the world. This has been made possible by conceptual differentiation empowered by language, which overtook emotional synthesis. But the loss of synthesis has also lead to ambiguity of meanings and values. Current English language cultures face internal crises, uncertainty about meanings and purposes. Many people cannot cope with diversity of life. Future research in psycholinguistics, anthropology, history, historical and comparative linguistics, and cultural studies will examine interactions between languages and cultures. Initial experimental evidence suggests emotional differences among languages consistent with our hypothesis (Guttfreund, 1990; Harris, Ayçiçegi, & Gleason, 2003).

Neural mechanisms of grammar, language sound, related emotions-motivations, and meanings hold a key to connecting neural mechanisms in the individual brains to evolution of cultures. Studying them experimentally is an ongoing research direction. The following sections develop mathematical models based on existing evidence that can guide this future research.

4.9 Hierarchical Evolving Systems, the Beautiful and Sublime

Influence of language emotionality on evolution of languages and cultures can be studied by simulating societies of intelligent agents. Agent's minds and communications can be modeled by using mathematical models of cognition, language, and their interactions in sections 3.7, 4.2, 4.3. This is a project for future research, which will take several books. Such large-scale simulations should be guided by simpler models that could be studied by simpler means. We explore such simpler models in this and following sections.

This section summarizes mathematical models of the mind mechanisms corresponding to the discussion in the previous section. These models are based on the available experimental evidence and theoretical development by many authors summarized in (Perlovsky, 1987; 1994; 1997; 1998; 2000; 2006a,b,c; 2007b; 1009; Perlovsky, Plum, Franchi, Tichovolsky, Choi, & Weijers, 1997) and it corresponds to recent neuro-imaging data (Bar et al, 2006; Franklin et al, 2008).

4.9.1 Hierarchical Model of Cognition

Here we consider steps toward extending a mathematical model of cognition considered in chapters 2 and 3 to the hierarchy of the mind discussed in section 4.1. This is a problem of immense complexity; it should be appropriately studied by simulating societies of intelligent agents, and here our goal is to derive approximate equations, which could guide this future exploration. We start with equation 2.1-5, and modify it to account for various penalty terms that we have ignored previously, and that should be accounted for, when considering the entire hierarchy of the mind.

$$L = \prod_{n \in N} \sum_{m \in M} r(m)\, l(n|m)\, pe(N,M)\, o(N,M)\, v. \qquad (4.9.1)$$

Here, as in (2.1-5) $l(n|m)$ is a conditional similarity of a bottom-up signal in pixel n given that it originated from the top-down concept-model m. Function $pe(N,M)$, penalizes for the number of parameters in models, $o(N,M)$ penalizes for the number of computations, and v is Vapnik's penalty function (Vapnik, 1998) discussed now in some details. The penalty for the number of parameters is necessary, because given a large number of parameters, virtually any model can describe any set of bottom-up signals. These models, however, would not have much predictive value for describing new signals. We have discussed this problem in section 2.4. Similar is the role of Vapnik's penalty function, which penalizes not for the number of parameters, but directly for the flexibility of the set of all models. If a given set of models can describe any set of bottom-up signals, the models are "too flexible," they would have no predictive power and this property is penalized by Vapnik's penalty. The number of computations has to be penalized because the mind-brain has to function effectively in real time, and the number of computations is a costly reserve. Specific functional shapes of these functions we consider later.

As discussed, (4.9-1) describes the knowledge instinct operating at a single level of the mind hierarchy. Some neural modelers like to emphasize that the mind-brain is not a strict hierarchy; it involves cross-interaction among multiple layers. For simplicity we use the word hierarchy. To describe the hierarchy, we denote a single-layer similarity (4.9-1) and all characteristics of this layer by index $h = 1, \ldots H$. The total similarity, specifying the instinct for knowledge for the entire hierarchy,

$$L = \prod_{h} L_h \; ; \qquad (4.9.2)$$

Mathematical models connecting this neural brain modeling to cultural evolution can proceed by simulating societies of interacting agents, each one satisfying its instinct for knowledge, and in addition, communicating through language. Here, we derive simplified expressions for similarity averaged over a population, so that maximization of similarity (4.9-2) could be studied analytically. Averaging over population is equivalent to studying cultures, rather that individual minds.

Similarity (4.9-2) determines the dynamics of multi-agent societies not unlike Lagrangian in physics determines the behavior of complex systems. Correspondingly, we use a technique inspired by mean field theories in physics, which have been developed for studying complex systems by substituting certain stochastic parameters in Lagrangian by their average values.

4.9.2 The Mean Field Hierarchical Dynamics

Considering (4.9-1) as a layer in (4.9-2), bottom-up signals are substituted by activated models at a lower layer, $N_h = M_{h-1}$. We take parameter penalty function, which exactly compensates for the effect of multiple parameters in the asymptotical regime, when the number of bottom-up signals is large (Akaike, 1974), this asymptotic regime, $N_h >> M_h$, is expected to be appropriate because only a small number of bottom-up signals, M_{h-1}, are organized into meaningful concepts M_h,

$$pe(h) = exp\{ -p^* M_h / 2 \}. \qquad (4.9.3)$$

Here p is an average number of parameters per model (the layer index h is sometimes omitted for brevity). A penalty for the number of computations, $o(h) = 1 / (number\ of\ operations)$; the number of operations is proportional to the product of bottom-up and top-down signals,

$$o(h) = c2(h) / (M_{h-1} * M_h * p), \tag{4.9.4}$$

were $c2(h)$ is a constant. We repeat, at every layer h, only a tiny part of all possible combinations of bottom-up signals, M_{h-1}, are organized into meaningful concepts M_h; a majority of combinations do not have any meaning; they are assigned to a "clutter" model ("or everything else"). The clutter model is homogeneous (does not depend on input data, and is only characterized by its proportion of signals, or rate, r_c. Concept-model rates at layer h, $r(m, h)$, are proportions of M_{h-1} signals associated with model $m(h)$; they are replaced by their average values, r_h. According to the rate normalization (2.1-4),

$$\sum_{m \in M(h)} r(m,h) + r_c = 1, \text{ or } M_h * r_h + r_c = 1. \tag{4.9.5}$$

Psychologically, at level h, $M_h r_h$ is proportional to the total amount of knowledge, therefore we introduce a notation, $K_h = M_h r_h$; correspondingly, clutter is proportional to the "unknown". Equation (4.9-5) is equivalent to

$$r_c = 1 - K_h; \quad K_h = M_h r_h. \tag{4.9.6}$$

These definitions correspond to normalizing the total number of known and unknown knowledge at level h, $r_c + K_h = 1$.

Vapnik's penalty penalizes "too flexible" models, which can explain everything. In a simplified way, it penalizes for K_h -> 1. Accordingly, as an approximation, we define it as

$$v(h) = exp\{ -v/(1-K_h) \}. \tag{4.9.7}$$

The average value of $l(m|n)$ can be computed as follows. For a large number of data, any functional shape of conditional similarities $l(m|n)$ (sections 3.1-3.7), can be modeled by a Gaussian function of $\Delta\mathbf{X}$, deviations of data, $\mathbf{X}(n)$, from the model m, $\mathbf{M}_m$, with covariance matrix $\mathbf{C}$, with dimensionality equal to the number of model parameters, p,

$$l(m|n) = (1/2\pi)^{p/2} det(\mathbf{C})^{-p/2} \, exp\{-(\Delta\mathbf{X}\mathbf{C}^{-1}\Delta\mathbf{X}/2)\}. \tag{4.9.8}$$

For evaluating of an average value of $l(m|n)$ we assume that concept recognition is nearly perfect, so $l(m|n) \sim \delta_{mn}$. This is appropriate because the concept learning is guided by language, and majority of concepts are well separated—otherwise we will not be able to function—confusion among concepts, when learning new ones, is a relatively rare events among the large amount of concepts. The average value of $det(\mathbf{C})$ is substituted with σ^{2p}, σ being an average standard deviation. In the exponent, $<\Delta\mathbf{X}\, \Delta\mathbf{X}> = \mathbf{C}$, and

$$< - \Delta\mathbf{X}\, \mathbf{C}^{-1}\, \Delta\mathbf{X} / 2 > = -1/2\, Tr(\mathbf{1}) = -p/2. \tag{4.9.9}$$

So the average value of conditional similarity,

$$<l(m|n)> = (1/2\pi)^{p/2}\, (1/\sigma^p)\, exp\{ - p/2 \}\delta_{mn}. \tag{4.9.10}$$

Psychologically, this partial similarity models an emotional certainty that data n originates from concept m. We denote it

$$E = <l(m|n)>. \tag{4.9.11}$$

Emotionality of knowledge, as discussed, depends on emotionality of language: language drives details vs. generality of cognitive models and determines ranges of σ and E. Detailed mathematical models of this interaction suitable for modeling of the hierarchical dynamics is a matter of future research.

Combining the above, a mean value of a layer h similarity,

$$L_h = [1-K_h + K_h E_h]^{M(h-1)} exp[-p_h M_h/2 - v/(1-K_h)]o(h). \tag{4.9.12}$$

Here K and M characterize the breadth and differentiation of knowledge, whereas E characterizes emotional certainty about validity of knowledge. This mean-field expression for similarity, together with eq.(4.9-2) can be used now to derive equations governing hierarchical dynamics of the knowledge instinct, which defines emotional and knowledge-oriented "spiritual" individual ontological development—on average—or more appropriately, social dynamics of cultural evolution. This dynamics according to the ideas of the knowledge instinct (KI) is given by the standard procedure of defining temporal derivatives along the gradient of similarity or KI. This dynamics leads to evolution that satisfies KI,

$$dE_h/dt = \delta dL/dE_h = \delta L^* d(ln L_h)/dE_h = \delta L^* M_{h-1}{}^*K_h/[1-K_h + K_h E_h], \tag{4.9.13}$$

$$dK_h/dt = \delta dL/dK_h = \delta L^* \{M_{h-1}{}^*(E_h-1)/[1-K_h + K_h E_h] - v/(1-K_h)^2\}, \tag{4.9.14}$$

$$dM_h/dt = \delta dL/dM_h = \delta L^* \{ln[1-K_{h+1} + K_{h+1}E_{h+1}] - p_h/2 -1/M_h\}, \tag{4.9.15}$$

where δ is a coefficient defining an evolutionary step and that would have to be determined empirically.

In addition to this knowledge-instinct driven dynamics, the hierarchy grows or shrinks depending on expansion or contraction of the number of general concepts at each layer. More general concepts move to higher levels of the hierarchy, and vice versa. The generality of a concept is determined by its standard deviation, related to emotionality, eqs.(4.9-11,4.9-12). Detailed description of this part of hierarchical dynamics would require accounting for standard deviations varying from a typical value for each layer. Modeling this process in the future will account for interaction between language and cognition, and for the distribution of standard deviations, σ_h, at every layer. As discussed this future research should be addressed by simulating societies of intelligent agents. Here our goal is a qualitative analysis aimed at deriving simpler equations. Taking a simple assumption that the distribution of σ_h at every layer is similar, would lead to a number of models moving between layers proportional to the number of models at each layer

$$dM_h/dt \sim (M_{h+1} - 2M_h + M_{h-1}), \tag{4.9.16}$$

Since the number of concepts at lower layers is much larger than at higher ones, this equation might lead to a growing hierarchy; however, combining this dynamics with eq.(13,14,15) would require a detailed numerical study.

Maximizing eq.(4.9-1) even for a single layer in case of few specific objects is a highly complex problem, rarely solved (like in chapter 3). Deriving relatively simple equations (4.9-13) through (4.9-16) for the evolution of the entire hierarchy is a major step. Nevertheless, this section "glossed over" mechanisms of interaction between cognition and language. The future research will derive the necessary more comprehensive equations, and explore their solutions; this might take more than one book. In the following sections we use the above equations as an intuitive, qualitative guide for deriving simpler equations, which can be explored within the limits of this book.

4.10 Evolution of Cultures

Qualitative examination of eqs.(4.9-13,14,15) indicates two mechanisms with opposing tendencies: differentiation and synthesis. Differentiation drives creation of a large number of detailed models, whereas synthesis unifies these detailed models at higher hierarchical levels. 3 regimes or solution types can be identified. The first, $E, K, M \sim 0$ and their time derivatives are also near 0. This could be characterized as primordial consciousness. The second, $K \sim< 1$, $E >> 1$; time derivatives are near 0. This could be characterized as traditional consciousness, there is no strivings for unknown, everything seems understood and fixed, emotional certainty in this limited knowledge is high. The third, is a knowledge-acquiring consciousness, with $(1\text{-}K) \sim KE$ and a non-trivial dynamics.

For detailed examination we derive simplified equations for this process in correspondence with properties of the above equations and their psychological interpretations discussed in previous sections. This would lead to approximate descriptions of cultural evolutions and guide future research. Let us summarize these previous discussions.

The hierarchical dynamics of the knowledge instinct manifests in two opposing tendencies, differentiation and synthesis. Differentiation satisfies KI by developing more specific and detailed models at lower levels; it acts at each single layer and drives creation of concrete, specific concepts—in other words, it drives top-down processes in the hierarchy, developing detailed concept-models. Synthesis satisfies KI by developing more general unifying models; it drives bottom-up processes in the hierarchy, it drives creation of general concept-models at a higher level, unifying differentiated models at lower levels. Differentiation is necessary for detailed understanding of the surrounding. Synthesis creates unified meanings of diverse experience; it is necessary for concentrating will and directing it to the most important goals.

Differentiation and synthesis are in complex relationships, at once symbiotic and antagonistic. Synthesis creates emotional value of knowledge, it unifies language and cognition, creates psychological conditions for differentiation; it leads to spiritual inspiration, to active creative behavior leading to fast differentiation, to creation of knowledge, to science and technology. At the same time, a "too high" level of synthesis, high emotional values of concepts stifles differentiation. Everyone has high-value emotional concepts related to a favorite football team, or political party, or family. Analyzing-differentiating these models is psychologically difficult because of strong emotions involved. When most of models in the entire culture are strongly emotional, differentiation and accumulation of knowledge stagnates, as in traditional cultures.

Depending on parameter values in the above equations, synthesis may lead to growth of general concept-models and to growth of the hierarchy. This is counterbalanced by differentiation. Differentiation leads to the growth of the number of concepts approaching "precise knowledge about nothing" ($E \to \infty$, $\sigma \to 0$, $r \to 0$). In the knowledge-acquiring regime the growth of synthesis is limited psychologically: emotions of the knowledge instinct satisfaction, when "spread" over large number of concepts, cannot sustain growing number of concepts, M. This is well known in many engineering problems, when too many models are used: everything can be explained, but this explanation has no predictive power. Akaike and Vapnik penalty functions, eqs.(4.9-3, 4.9-7), counterweigh, and the number of models falls. Thus, whereas emotional synthesis creates a condition for differentiation (high emotional value of knowledge, efficient dual model connecting language and cognition, large E, growth of K and M), conceptual differentiation undermines synthesis (value of knowledge, E, and its diversity, M, fall). This interaction can be modeled by the following equations:

$$dM/dt = aM\,G(S),\quad G(S) = (S - S_0)\,exp(-(S-S_0)/S_1),$$

$$dS/dt = -b\,M + d\,H,$$

$$H(t) = H_0 + e{*}t. \qquad (4.10.1)$$

Here, t is time, M is a number of concepts (differentiation), S models synthesis, H is a number of hierarchical levels; a, b, d, e, S_0 and S_1 are constants. Differentiation, M, grows proportionally to already existing number of concepts, as long as this growth is supported by synthesis, while synthesis is maintained at a "moderate" level, $S_0 < S < S_1$. "Too high" level of synthesis, $S > S_1$, stifles differentiation by creating too high emotional value of concepts. Synthesis, S, is related to emotion, E, but the detailed relationship will have to be established in future research by detailed analysis of equations (4.9-13) through (4.9-16). Synthesis, S, grows in the hierarchy, along with a number of hierarchical levels, H. By creating emotional values of knowledge, it sustains differentiation, however, differentiation, by spreading emotions among a large number of concept-models destroys synthesis. Analysis of hierarchical dynamics H qualitatively from eqs. (4.9-13) through (4.9-16) is difficult, so instead we just consider a period of slow growth of the hierarchy H. At moderate values of synthesis, solving eqs.(4.10-1) yields a solution in Fig. 4.10-1. The number of concepts grows until certain level, when it results in reduction of synthesis; then the number of models falls. As a number of models falls, synthesis grows, and the growth in models resumes. The process continues with slowly growing, oscillating number of models Oscillations affecting up to 80% of knowledge indicate internal instability of this knowledge-accumulating culture. Significant effort was extended to find solutions with reduced oscillations, however, no stable knowledge-acquiring solution was found based on eqs.(4.10-1). This discussion is continued below (Fig.4.10-3).

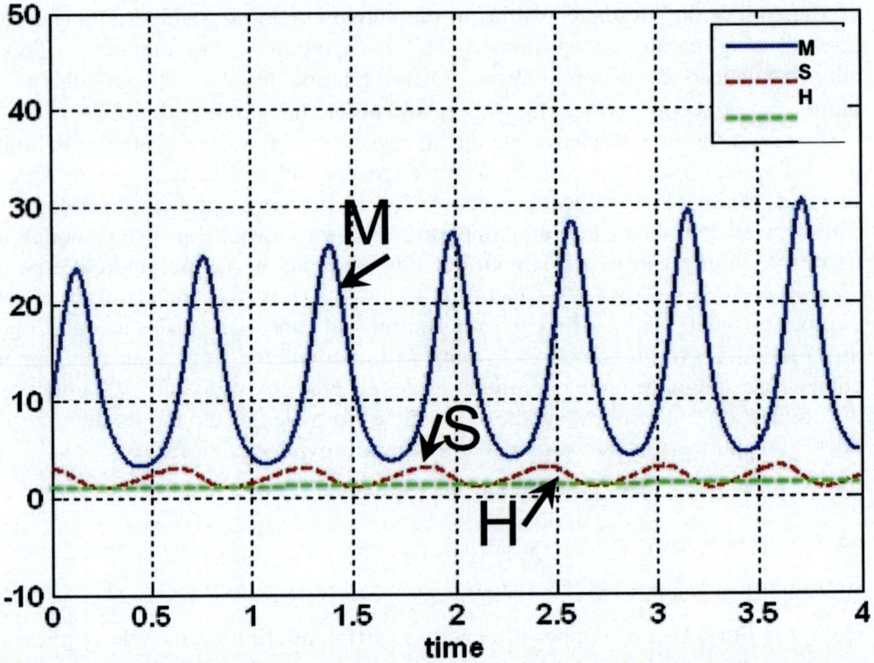

Fig. 4.10-1 Evolution of culture at moderate values of synthesis oscillates: periods of flourishing and knowledge accumulation alternate with collapse and loss of knowledge ($a = 10, b = 1, d = 10, e = 0.1, S_0=2, S_1=10$, and initial values $M(t=0) = 10, S(t=0) = 3, H_0 = 1$; parameter and time units are arbitrary). In long time the number of models slowly accumulates; this corresponds to slowly growing hierarchy.

Another solution corresponds to initially high level of synthesis, Fig. 4.10-2. Synthesis continues growing whereas differentiation levels off. This leads to a more and more stable society with high synthesis, in which high emotional values are attached to every concept, however, differentiation stagnates.

These two solutions of eqs.(4.10-1) can be compared to Humboldt's (1836/1967) characterization of languages and cultures. He contrasted inert objectified "outer form" of a language vs. subjective, culturally conditioned, and creative "inner form." Humboldt's suggestion continues to stir linguists' interest today, yet seems mysterious and not understood scientifically.

Our analysis suggests the following interpretation of Humboldt's thoughts in terms of neural mechanisms. His "inner form" corresponds to the integrated, moderately emotional neural dual model. Contents of cognitive models are being developed guided by language models, which accumulate cultural wisdom. "Outer form" of language corresponds to inefficient state of neural dual model, in which language models do not guide differentiation of the cognitive ones. This might be due to either too strong or too weak involvement of emotions. If emotional involvement in cognition or language is too weak, learning does not take place because motivation disappears. If emotional involvement is too strong, learning does not take place because old knowledge is perceived as too valuable, and no change is possible. The first case might be characteristic of low-inflected languages, when sound of language changes "too fast," and emotional links between sound and meanings are severed. The second case might be characteristic of "too strongly" inflected languages, in which sound changes "too slowly" and emotions are connected to meanings "too strongly;" this could be a case of Fig. 4.10-2. A brief look at cultures and languages certainly points to many examples of this case: highly inflected languages and correspondingly

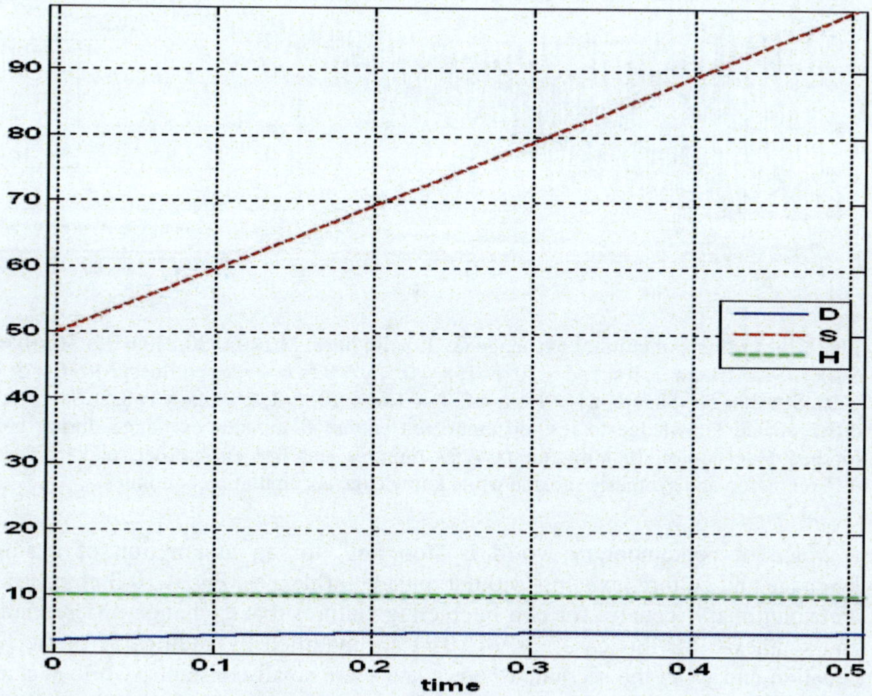

Fig. 4.10-2 Evolution of highly stable, stagnating society with growing synthesis. High emotional values are attached to every concept, while knowledge accumulation stops ($M(t{=}0){=}\ 3,\ H_0 = 10,\ S(t{=}0) = 50,\ S_0 = 1,\ S_1 = 10,\ a = 10,\ b = 1,\ d = 10,\ e{=}1$).

"traditional" stagnating cultures. Which of these correspond to Fig. 4.10-2 and the implied neural mechanisms? What it means quantitatively: "too fast" or "too slow," and which cultures and languages correspond to which case will require further psycholinguistic and anthropological research.

The integrated dual model assumes "moderate" emotional connection between language and cognitive models, which fosters the integration and does not impede it. Humboldt suggested that this relationship is characteristic of inflectional languages (such as Indo-European), inflection provided "the true inner firmness for the word with regard to the intellect and the ear" (according to our analysis we would say "concepts and sounds-emotions"). The integrated dual model assumes a moderate value of synthesis, Fig. 4.10-1, leading to interaction between language and cognition and to accumulation of knowledge. This accumulation, however, does not proceed smoothly; it leads to instabilities and oscillations, possibly to cultural calamities; this characterizes significant part of European history from the fall of Roman Empire to recent times.

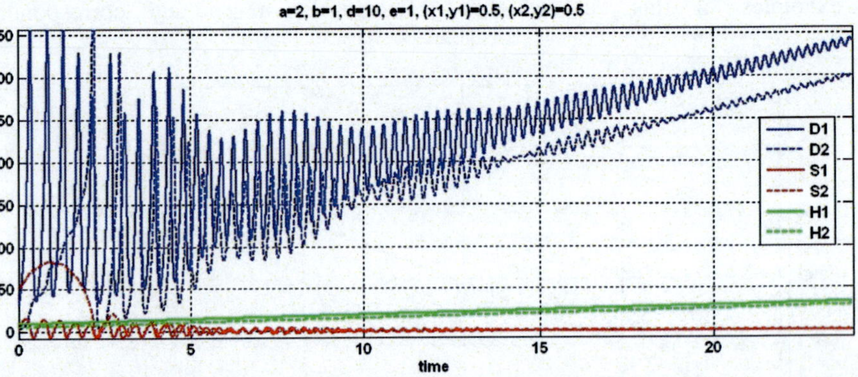

Fig. 4.10-3 Effects of cultural exchange (k=1, solid lines: M(t=0)= 30, H0 = 12, S(t=0) = 2, S0 = 1, S1 = 10, a = 2, b = 1, d = 10, e=1, x = 0.5, y = 0.5; k=2, dotted lines: M(t=0)= 3, H0 = 10, S(t=0) = 50, S0 = 1, S1 = 10, a = 2, b = 1, d = 10, e=1, x = 0.5, y = 0.5). Transfer of differentiated knowledge to less-differentiated culture dominates exchange during t < 2 (dashed blue curve). In long run (t > 5), cultures stabilize each other, and swings of differentiation and synthesis subside while knowledge accumulation continues.

Much of contemporary world is "too flat" for an assumption of a single language and culture, existing without outside influences. Fig. 4.10-3 emonstrates an evolutionary scenario for two interacting cultures that exchange differentiation and synthesis; for this case eqs. (4.10-1) are modified by adding xM to the first equation and yS to the second, where x and y are small constants, while M and S were taken from the other culture. The first and second cultures initially correspond to Figs. 4.10-1 and 4.10-2 correspondingly. After the first period when the influence of the first culture dominates, both cultures stabilize each other, both benefit from fast growth and reduced instabilities.

4.11 Emotional Sapir-Whorf Hypothesis

Misunderstanding among cultures possibly is the most important challenge for the 21st century. To improve cultural understanding, similarities and differences among cultures have to be scientifically analyzed.

4.11.1 Determinants of Cultural Evolution

As we discussed in section 4.3, language is a most important mechanism of transmitting information through generations. Language and cognition are in constant interaction. This interaction is not symmetrical. Language drives cognition and not the other way around. Let us quickly recollect the argument. Cognition is grounded in perception only at the lower levels of the mind hierarchy, at the level of perceptual features and objects, which can be directly perceived. Higher abstract thoughts, which set human thinking apart from the animal world, cannot be directly perceived. Every human child acquires higher, abstract level cognition due to guidance by language. Language is grounded in the surrounding language at all hierarchical levels. At all levels language exist in the surrounding culture "ready-made." This is the reason that kids learn language by 5 years of age, but it takes the rest of life to acquire cognitive models, which include adult understanding. This adult understanding includes cognitive models of the world, self, and behavior, corresponding to the surrounding culture contemporary with individuals. And most individuals during their lifetime learn only small part of knowledge existing in culture and language. Limits of knowledge stored in language determine what most of people can learn in their lifetime. Beyond this limit begins a slow process of knowledge development, which speed is measured not in decades and lifetimes, but in tens and hundreds of lifetimes.

The most important contents of languages are conceptual contents, the diversity of concepts used to understand the world. In the 1930s Benjamin Whorf and Edward Sapir analyzed influence of conceptual contents of languages on the development of cultures. There was a long predating linguistic and philosophical tradition, which emphasized influence of language on cognition (Bhartrihari, IVCE/1971; Humboldt, 1836/1967; Nietzsche, 1876/1983), nevertheless, the idea of language affecting culture is often referenced as Sapir-Whorf hypothesis (SWH). Linguistic evidence in support of this hypothesis concentrated on conceptual contents of languages. For example, words for colors influence color perception (Roberson, Davidoff, & Braisbyb, 1999; Winawer, Witthoft, Frank, Wu, Wade, & Boroditsky, 2007). The idea of language influencing cognition and culture has been criticized and "fell out of favor" in the 1960s (Wikipedia, 2009a) due to a prevalent influence of Chomsky's ideas emphasizing language and cognition to be separate abilities of the mind (Chomsky, 1965). Recently SWH again attracted attention of linguists.

According to our previous analysis we would like to emphasize that emotional contents of languages could be no less important than conceptual contents. Conceptual contents are stored in words and to extent could be borrowed among languages. But emotional contents are stored in sounds determined by grammar, which cannot be borrowed. Sounds of languages, and therefore emotional contents, relations between conceptual contents and emotionalities of languages differ more than concepts alone. Therefore, summarizing previous discussions we propose the Emotional SWH (ESWH), a hypothesis suggesting that language emotions and grammars influencing language sounds, determine cultural differences.

To correctly interpret SWH, ESWH, and the discussed cultural differences, it is essential to remember that these differences determine evolution and contents of cultures. They are not personal. Every individual can learn many languages and acquire conceptual and emotional contents of many cultures.

In the milieu defined by Chomsky's assumed independence of language and cognition the Sapir-Whorf hypothesis (SWH) has steered much controversy:

"This idea challenges the possibility of perfectly representing the world with language, because it implies that the mechanisms of *any* language condition the thoughts of its speaker community" (Wikipedia, 2008).

The fact that Wikipedia seriously considers a naïve view of "perfectly representing the world" as a scientific possibility is indicative of a problematic state of affairs, "the prevalent commitment to uniformitarianism, the idea that earlier stages of languages were just as complex as modern languages" (Hurford, 2008). With the development of cognitive and evolutionary linguistics diversity of languages are considered in their evolutionary reality, and identifying neural mechanisms of language evolution and language-cognition interaction is coming in demand.

4.11.2 Predictive Cultural Models

Previous sections described models of cultures, which explain cultural differences and similarities and predict directions of cultural evolution. These models are approximate in terms of psychological effects accounted for and in terms of mathematical accuracy. More accurate models should simulate societies of intelligent agents with cognitive and language abilities as discussed in section 4.3, agents that interact using a language-like ability, form societies, and accumulate knowledge, similar to human cultures. These simulated intelligent agents should account for human emotional abilities. The principal emotional ability is that for musical emotions, which we consider in section 4.11.

Future mathematical-theoretical research should address continuing development of both mean-field and multi-agent simulations, connecting neural and cultural mechanisms of emotions and cognition and their evolution mediated by language. The knowledge instinct theory should be developed toward theoretical understanding of its differentiated forms explaining multiplicity of aesthetic emotions in language prosody and music (Perlovsky, 2006d; 2008). This theoretical development should go along with experimental research clarifying neural mechanisms of the knowledge instinct (Levine & Perlovsky, 2008; Bar et al, 2006) and the dual language-cognitive model, (Perlovsky, 2009).

4.11.3 Experimental Evidence and Future Research

Models of cultural evolution discussed above are but an initial step in this line of research. Nevertheless concrete predictions are made for relations between language grammars and types of cultures. These predictions can be verified, coefficients in eqs.(4.10-1) can be measured in psycholinguistic laboratories. Initial evidence indicates different emotionalities in different languages consistent with expected for the language grammars, more inflective languages are more emotional (Guttfreund 1990; Harris, Ayçiçegi, & Gleason 2003).

Experimental results on neural interaction between language and cognition (Franklin et al, 2008; Simmons et al, 2008) support the mechanism of the dual model. They should be expanded to interaction of language with emotional-motivational, voicing, behavioral, and cognitive systems.

Prehistoric anthropology should evaluate the hypothesis that the primordial system of fused conceptual cognition, emotional evaluation, voicing, motivation, and behavior differentiated at different prehistoric time periods. Are there data to support this hypothesis, can various stages of prehistoric cultures be associated with various neural differentiation stages? Can different humanoid lineages be associated with different stages of neural system differentiation? What stage of neural differentiation corresponds to Mithen's hypothesis about singing Neanderthals (Mithen, 2007)? Psychological social and anthropologic research should go in parallel documenting various cultural evolutionary paths and correlations between cognitive and emotional contents of historical and contemporary cultures and languages.

Proposed correlation between grammar and emotionality of languages can be verified in direct experimental measurements using skin conductance and fMRI neuro-imaging. Emotional version of Sapir-Whorf hypothesis should be evaluated in parallel psychological and anthropological research. More research is needed to document cultures stagnating due to "too much" emotionality of languages; as well as crises of lost values due to "low" emotionality of language (e.g. in English-speaking countries). Hieroglyphic writing (Chinese) separates sounds and meaning; how this affects functioning of the dual model? How interaction of emotional and conceptual contents of cognition are affected by tonal languages (in most European languages pitch or tone of voice indicates emotion, and bears no separate conceptual meaning; in tonal languages, such as Chinese, pitch is also used to communicate conceptual meanings).

4.12 Music: Its Function in Cognition and Evolution

4.12.1 An Unsolved Mystery

Music is a mystery. Functions and origins of music have challenged philosophical thought for thousands of years. Aristotle listed the power of music among the unsolved problems (Aristotle, IV BCE/1995, p.1434). According to Darwin (1871), it "must be ranked amongst the most mysterious (abilities) with which (man) is endowed." Recently research resurged in relating music to emotions (Juslin & Sloboda 2001). The suggestion that music and emotions are linked opens more questions than answers: how music expresses or creates emotions, are these emotions similar or different from other emotions, what is their function? "Music is a human cultural universal that serves no obvious adaptive purpose, making its evolution a puzzle for evolutionary biologists" (Masataka, 2008). Kant (1790), who so brilliantly explained the epistemology of the beautiful and the sublime, could not explain music: "(As for) the expansion of the faculties… in the judgment for cognition, music will have the lowest place among (the beautiful arts)… because it merely plays with senses." Pinker (1997) follows Kant, suggesting that music is an "auditory cheesecake," a byproduct of natural selection that just happened to "tickle the sensitive spots." In 2008, Nature published a series of essays on music. Their authors agreed that music is a cross-cultural universal, still "none… has yet been able to answer the fundamental question: why does music have such power over us?" (Editorial, 2008). "We might start by accepting that it is fruitless to try to define 'music'." (Ball, 2008). These are just a sampling of quotes from accomplished scientists.

Here we present a theory or hypothesis based on previous arguments in this chapter suggesting that music serves the most important and concrete function in evolution of the mind and cultures. We discuss this function, neural mechanisms, and suggest experimental verification of this hypothesis.

4.12.2 2,500 Years of Western Music and Pre-scientific Theories (from Pythagoras to the 18th c.)

Pythagoras described the main harmonies as whole-number ratios of sound frequencies about 2,500 years ago. He saw this as a connection of music to celestial spheres, which also seemed governed by whole numbers (James 1995). In the pre-scientific era, musical thoughts were led by composer's practice and philosophical thoughts followed behind. The tremendous potency of music to affect consciousness, to move people's souls and bodies since time immemorial was ambivalently perceived. Ancient Greek philosophers saw human psyche as prone to dangerous emotional influences and "proper" music was harmonizing human psyche with reason. Plato wrote about idealized imagined music of the Golden Age of Greece: "... (Musical) types were… fixed… Afterwards… an unmusical license set in with the appearance of poets… men of native genius, but ignorant of what is right and legitimate… Possessed by a frantic and unhallowed lust for pleasure, they contaminated… and created a universal confusion of forms… So the next stage… will be… contempt for oaths… and all religion. The spectacle of the Titanic nature… is reenacted; man returns to the old condition of a hell of unending misery." (Plato 4c. BCE).

Plato's prediction has come to pass many times over, man has returned to a hell of unending misery. But is a wrong music to be blamed every time?

The same appeal to reason as a positive content of music we find 800 years later in Boethius (5c.) "…what unites the incorporeal existence of reason with the body except a certain harmony, and, as it were, a careful tuning of low and high pitches in such a way that they produce one consonance?" (see in Weiss & Taruskin 1984; unreferenced quotes in this section refer to this book). According to foremost thinkers in the 4th and 5th centuries (including St. Augustine) the mind was not strong enough to be reliably in charge of senses and unconscious urges. Differentiation of emotions was perceived as a danger.

Only with the beginning of the Renaissance (13-14th c.), for the first time since antiquity the European man felt the power of rational mind separating from collective consciousness (that is, from received cultural rules). The millennial tradition of music understanding was changing. For twelve centuries, Plato, Boethius, and Erigena (from 4th c. BCE to 9th c. AD) saw the positive content of music in its relations to objective 'motion of celestial spheres' and to God-created laws of nature. This changed by the 13th century: The music was now understood as related to listeners, not to celestial spheres. J Groceo (14th.c.) wrote: Songs for "average people… relate the deeds of heroes… the life and martyrdom of various saints, the battles…"; songs for kings and princes "move their souls to audacity and bravery, magnanimity and liberality..." Human emotions, the millennial content of music, were appreciated theoretically.

Whereas music appealed to emotions since time immemorial, a new and powerful development toward stronger and more diverse emotionality started during the Renaissance. It came with the tonal music developed for 500 years from the 15th to 19th c. with a *conscious* aim of appealing to musical emotions. (Tonality is the system of functional harmonic relations, governing most of the Western music. The tonal music is organized around tonic, a privileged key to which melody returns. Melody leads harmony, and harmony in turn leads melody. A melodic line feels closed, when it comes to rest on (resolved in) tonic. Emotional tension ends and a psychological relaxation is felt in the final move on to the tonic, to a resolution in a "cadence".

Creating emotions was becoming the primary aim of music. Composers strived to imitate speech, the embodiment of the passions of the soul. At the same time conceptual content of texts increased, "the words (are to be) the mistress of the harmony and not its servant," wrote Monteverdi at the beginning of the 17th c. This became the main slogan of the new epoch of Baroque music. The opera music was born in Italy at that time.

The nature of emotions became a vital philosophical issue. Descartes attempted a scientific explanation of passions (1646). He rationalized emotions, explaining them as objects and relating to physiological processes. "Descartes descriptions of the physiological processes that underlay and determined the passions were extremely suggestive to musicians in search of technical means for analogizing passions in tones."

Based on Descartes' theory, Johann Mattheson (1739) formulated a theory of emotions in music, called "The Doctrine of the Affections." Emotions "are the true material of virtue, and virtue is naught but a well-ordered and wisely moderate sentiment." Now the object of musical imitation was no longer speech, the exterior manifestation of emotions, but the emotions themselves."

Beginning from this time musical theory did not just trail musical practice but affected it to significant extent. Descartes and Mattheson understood emotions as monolithic objects. This simplified understanding of emotions soon led to deterioration of opera into a collection of airs, each expressing a particular emotion ("opera seria" or serious opera); the Monteverdi vision of opera as integrated text, music, and drama was lost. In the middle of the 18th c. Calzabigi and Gluck reformed opera back to the Monteverdi vision and laid a theoretical foundation for the next 150 years of opera development.

As we discuss later, music is different from other arts in that it affects emotions directly (not through concepts-representations). This clear scientific understanding of the differences between concepts and emotions did not exist. Nevertheless, an idea of music as expression, differentiating (creating new) emotions, was consciously formulated in the second half of the 18th c. (C. Avison, 1753 and J. Beattie, 1778). This idea of music as expression of emotions led to a fundamental advancement in understanding music as the art differentiating (creating new) emotions; it related the pleasures of music sounds to the 'meaning' of music. T. Twining (1789) emphasized an aspect of music, which today we would name conceptual indefiniteness: musical contents cannot be adequately expressed in words and do not imitate anything specific. "The notion, that painting, poetry and music are all Arts of Imitation, certainly tends to produce, and has produced, much confusion… and, instead of producing order and method in our ideas, produce only embarrassment and confusion." (in W&T, pp. 293-294).

Yet understanding the nature of emotions remained utterly confused: "As far as (music) effect is merely physical, and confined to the ear, it gives a simple original pleasure; it expresses nothing, it refers to nothing; it is no more imitative than… the flavor of pineapple." Twinning expresses here correct intuition (music is not an imitation), but he confuses it with a typical error. Pleasure from musical sounds is not physical and not confined to the ear, as many have thought. As discussed later, pleasure from music is an aesthetic (not bodily) emotion in our mind unlike, for example, the flavor of a pineapple which promises to our body enjoyment of a physical food. Even the founder of contemporary aesthetics, Kant (1790) had no room for music in his theory of the mind: "(As for) the expansion of the faculties which must concur in the judgment for cognition, music will have the lowest place among (the beautiful arts)… because it merely plays with senses." (Later we discuss a specific scientific reason preventing Kant from understanding the role of music in cognition). Even today, as discussed in section 2.3, the role of musical emotions and their interaction with cognition remain little known among musicologists; the idea of expression continues to provoke disputes, "embarrassment and confusion."

4.12.3 Whence Beauty in Sound?

A scientific theory of music perception began its development in the first half of the 19th century by Helmholtz's (1863) theory of musical emotions, summarized in this section. A pressed piano key or plucked string produces a sound with many frequencies. In addition to the main frequency F, the sound contains overtones or higher frequencies, 2F, 3F, 4F, 5F, 6F, 7F…, which sound quieter than F. The main tone corresponds to the string oscillating as a whole, producing F; on top of this, each part of a string (1/2, or 1/3 or 2/3...) can oscillate on its own. A synthesizer can produce a sound with a single frequency F; it sounds similar to the ear as a piano key with the same main frequency, but more 'mechanical'. If one produces the key F, and at the same time 2F (quieter), then an untrained ear hears it very similar to the piano key. If all overtones are added, the sound will match the piano key. The interval between F and 2F (double frequency) is called an octave. If F is "Do, first octave (256 Hz)", then 2F is the Do of the second octave.

Our ear almost does not notice an overtone exactly one octave higher, because the eardrum oscillates as a string in concordance with itself. For the same reason all exact overtones (2F, 3F, 4F…) are perceived in concordance with the main frequency F and among themselves. Because of the mechanical properties of the eardrum, two sounds with close frequencies (say, F and 0.95F) produce eardrum oscillations not only with the same frequencies but also with the difference of these frequencies (F - 0.95F = 0.05F). These low frequency oscillations are perceived as physically unpleasant (sounding "rough," and even painful, though at normal loudness they are barely perceived). Sounds with exactly same overtones (most loud ones) are perceived as concordant, agreeable, or 'mechanically pleasing'.

Is it possible to select concordant strings within octave, which main overtones equal 3F, 4F, 5F, 6F, 7F…? – Yes, it could be achieved by dividing these frequencies by 2: 3/2F, 4/2F, 5/2F, 6/2F, 7/2F... (say, by taking a string twice as long). These sounds are perceived by the ear as concordant with the main key (F) and among themselves. This concordance is not as good as among overtones of a single string, but much better than for random sounds. That is the reason for musical importance of the octave: Strings (or keys) separated exactly by an octave (double or half the frequency) have many of the exact same overtones and they sound concordant. Note, only the first of the above sounds, 3/2F, is within the fist octave (above F and below 2F); the rest are in the second octave and above. For a key to sound in the first octave and its overtones to coincide with those of Do, we may bring down each overtone by one more octave (or two, or three): 5/4F, 7/4F, 9/8F.

Notes obtained in this way, if we start with the three main overtones, make up the major scale, do, re, mi, fa, sol, la, ti – the white piano keys. They are perceived by the ear as concordant. The note fa, however, sounds more concordant if its first different overtone coincides with an overtone of do, 4F (therefore the fa key is chosen as fa = 4/3F). Concordance, or similarity of overtones, somewhat depends on the training of the ear, also not all overtones could be made completely concordant; therefore musical acoustics is not as simple as 2 x 2 = 4. Musical instruments were improved over thousands of years and they incorporate traditions and compromises. There are important differences among cultures in making musical instruments and tuning them. The most concordant keys do, fa, sol (or F, 4/3F, 3/2F) exist practically in all cultures (they are the most concordant because the first overtone of do is sol, and the first overtone of fa is do). Next four overtones closest in loudness and similarity add up to the major scale.

The minor scale is obtained if the three least concordant keys, mi, la, ti, are lowered by a half-tone (tone =1/7th of an octave), so that they are more concordant with the other less loud overtones. If one chooses the most concordant note among these three less concordant keys, the note la, the resultant 5-notes are called the pentatonic scale; it is used in Chinese music, in folk music of Scotland, Ireland, and in Africa.

The scale of an accurately tuned piano slightly differs from what is described above. The reason is that all overtones of all keys cannot coincide; scale based on overtones of do is not as well concordant with overtones of other keys. For example, an overtone of mi, similar to sol, is ¼ tone different from sol and sounds as a strong dissonance. For string instruments, such as a violin, it is not too important; a violinist can take the correct interval for each note, similarly a singer. But for keyboard instruments, like piano, this sound error is not correctable. Therefore, in the 16th century a well-tempered scale was developed, which divides an octave into 12 equal intervals (half-tones), so that errors in the main overtones are equally spread and all keys are slightly discordant. Concordant musical sounds are called consonances, and less concordant, dissonances. The exact meanings of these words change with culture.

Notwithstanding the Helmholtz's acoustic theory, there is a principled difference between the 'mechanical' agreeableness of concordant overtones and esthetic beauty of music. For example, the minor scale is esthetically interesting exactly due to its slight discordance. Therefore, Helmholtz's theory could not be accepted as a basis for musicology. Sound "concordance" depends to some extent on musical ear training, and musical theory is not as simple as two plus two. Musical instruments have been perfected for thousands of years and there are important differences among cultures. Acoustic properties of the human voice and ear do not guarantee that Mozart sounds 'naturally'. A single string sounds naturally in complete concordance with its overtones, but classical musical harmony used natural mechanisms of perception of consonances and dissonances for complex esthetic effects. Fundamental significance of Helmholtz's theory remained unclear because it was not connected to the aesthetic meaning of music.

Recent laboratory experiments confirmed that musical harmony is based on inborn mechanisms. Babies (beginning at 4-month) like consonant sounds and dislike dissonances. Evolution, it seems, used the mechanical properties of the ear for enhancing efficiency of the spoken communication channel. As a string made of inhomogeneous material sounds in discordance with itself, so does the human voice chord, when in stress or fear; it sounds discordant; and this discordance was perceived as unpleasant millions of years ago. In the basis of human voice communication, there are consonant combinations of sounds. These were gradually evolving into the emotionally filled melody of voice. Connection of voice sounds with the states of soul was inherent in our ancestors long before language began evolving toward conceptual content at the expense of the emotional one. Gradually, evolution shaped musical ability to create and perceive sound as something principally important, touching all of our being. This is why wolves howl at the Moon, whereas humans express such a diversity of emotions in sounds.

Another physical difficulty of Helmholtz's theory is that emotional perceptions of consonances and dissonances extends from contemporaneously sounding frequencies also to temporal sequences of tones, and this cannot be explained by beats of eardrum. Apparently, over millennia (or possibly over millions of years beginning in animals – this point might be contentious) neural mechanisms added to our perception of originally mechanical properties of ear. I'll add that Helmholtz did not touch the main question of why music is so important psychologically - this remained a mystery.

4.12.4 Current Theories of Musical Emotions

Current theories of musical emotions attempt to uncover this mystery by looking into its evolutionary origins. Justus and Hustler (2003) and McDermott and Houser (2003) review evidence for evolutionary origins of music. They emphasize that an unambiguous identification of genetic evolution as a source of music origins requires innateness, domain specificity for music, and uniqueness to humans (since no other animals make music in the sense humans do). The conclusions of both reviews are similar, i.e., "humans have an innate drive to make and enjoy music." There is much suggestive evidence supporting a biological predisposition for music. Certain basic abilities for music are guided by innate constraints.

Still, it is unclear that these constraints are uniquely human since they "show parallels in other domains." It is likely that many musical abilities are not adaptations for music, but are based on more general-purpose mechanisms. There are "some intriguing clues about innate perceptual biases related to music, but probably not enough to seriously constrain evolutionary hypothesis." "Available evidence suggests that the innate constraints in music are not specific to that domain, making it unclear, which domain(s) provided the relevant selection pressures." "There is no compelling reason to argue categorically that music is a cognitive domain that has been shaped by natural selection." In Nature's series of essays on music McDermott (2008) writes: "Music is universal, a significant feature of every known culture, and yet does not serve an obvious, uncontroversial function".

In commentaries to these reviews, Trainor (2008) argues that for higher cognitive functions, such as music, it is difficult to differentiate between adaptation and exaptation (structures originally evolved for other purposes and used today for music), since most such functions involve both "genes and experience." Therefore the verdict on whether music is an evolutionary adaptation should be decided based on advantages for survival. Fitch (2004) comments that biological and cultural aspects of music are hopelessly entangled, and "the greatest value of an evolutionary perspective may be to provide a theoretical framework." Livingstone and Thompson (2006) emphasize a multimodal nature of the engaging effect of musical experience and explore theories based on exaptations of "an earlier system of affective communication." It is therefore interesting, they suggest, exploring correlations between musicality and emotional intelligence. They emphasize human symbolic ability leading to art, including music and our capacity for "symbolic hierarchical systems."

Before reviewing other select authors, we would comment that the hypothesis advocated later in the current review corresponds to many of the suggestions and ideas in this section. In addition, we discuss a fundamental function of musical emotions in the evolution of language, mind, and culture, which is missing in other theories and which provides new directions to search for evolutionary mechanisms of music. The review relates to biological roots of music, to its origins in "an earlier system of affective communication," it bears on discussions of evolution vs. exaptation, and human symbolic ability.

Huron (1999) emphasizes that in the search for evolutionary origins of music it is necessary to look for complex multistage adaptations, built on prior adaptations, which might have evolved for several reasons. He discusses social reasons for music origins and lists several possible evolutionary advantages of music: mate selection, social cohesion, the coordination of group work, auditory development, developing auditory skills, refined motor coordination, conflict reduction, preserving stories of tribal origins. However, the list of possible uses of music by itself does not explain musical power over human psyche; does not explain why music and not some other, nonmusical activities have been used for these purposes.

Cross (2008a,b), Cross & Morley (2008) concentrate on evolutionary arguments specific to music. Cross integrates neuroscientific, cognitive, and ethno-musicological evidence and emphasizes that it is inadequate to consider music as "patterns of sounds" used by individuals for hedonic purposes. Music should be considered in the context of its uses in pre-cultural societies for social structuring, forming bonds, and group identities. A strong argument for evolutionary origins of music is its universality; music exists in all scientifically documented societies around the globe. Cross emphasizes that music possesses common attributes across cultures: it exploits the human capacity to entrain to social stimuli. He argues that music is necessary for the very development of culture. Cultural evolution is based on ability to create and perceive socio-intentional aspect of meaning. This is unique to human and it is created by music. Cross presents a three-dimension account of meaning in music, combining "biologically generic, humanly specific, and culturally enactive dimensions." Thus evolution of music was based on already existing in animal world biological and genetic mechanisms.

The capacity for culture (Cross, 2008b) requires transmission of information, but also the context of communication. Therefore "music and language constitute complementary components of the human communicative toolkit." The power of language is in "its ability to present semantically decomposable propositions." Language, because of its concreteness, on one hand enabled exchange of specific and complicated knowledge, but on the other hand could exacerbate oppositions between individual goals and transform an uncertain encounter into a conflict.

Music is a communicative tool with opposite properties. It is semantic, but in a different way than language. Music is directed at increasing a sense of 'shared intentionality.' Music's major role is social, it serves as an 'honest signal' (that is it "reveals qualities of a signaler to a receiver") with nonspecific goals. This property of music, "the indeterminacy of meaning or floating intentionality," allows for individual interactions while maintaining different "goals and meanings" that may conflict. Thus music "promotes the alignment of participants' sense of goals." Therefore Cross hypothesized that successful living in societies promoted evolution of such communication system.

Cross suggests that music evolved together with language rather than as its precursor. Evolution of language required a re-wiring of neural control over the vocal tract, and this control had to become more voluntary for language. At the same time a less voluntary control, originating in ancient emotional brain regions, had to be maintained for music to continue playing the role of 'honest signal.' Related differences in neural controls over the vocal tract between primates and humans were reviewed in Perlovsky (2005, 2006b, 2006e, 2007).

As juvenile periods in hominid lineages lengthened (altricialization), music took a more important role in social life (Cross & Morley 2008). The reason is that juvenile animals, especially social primates, engage in play, which prepares them to adult lives. Play involves music-like features, thus proto-musical activity has ancient genetic roots. Lengthening of juvenile periods was identified as possibly fundamental for proto-musical activity and for origin of music. Infant directed speech (IDS) has special musical (or proto-musical) qualities that are universal around the globe. This research was reviewed in Trehub (2003). She has demonstrated that IDS exhibits many similar features across different cultures. Young infants are sensitive to musical structures in human voice. Several researchers relate this sensitivity to the "coregulation of affect by parent and child" (Dissanayake, 2000), and consider IDS to be an important evolutionary mechanism of music origin. Yet, arguments presented later tell that IDS cannot be a full story of musical evolution.

Dissanayake (2008) considers music primarily as a behavioral and motivational capacity. Naturally evolving processes led to ritualization of music through formalization, repetition, exaggeration, and elaboration. Ritualization led to arousal and emotion shaping. This occurred naturally in IDS, in the process of mother-infant interaction, which in addition to specially altered voice involved exaggerated facial expressions and body movements in intimate one-to-one interaction. Infants 8 weeks old already are sensitive to this type of behavior, which reinforces emotional bonding. This type of behavior and the infants' sensitivity to it are universal throughout societies, which suggests an evolved inborn predisposition. Dissanayake further emphasizes that such proto-musical behavior has served as a basis for culture-specific inventions of ritual ceremonies for uniting groups as they united mother-infant pairs. The origins of music, she emphasizes, are multi-modal, involving aural, visual, and kinesic activity, which has occurred in social rather than solitary settings. She describes structural and functional resemblances between mother-infant interactions, ceremonial rituals, and adult courtship, and relates these to properties of music. All these, she proposes, suggest an evolved "amodal neural propensity in human species to respond—cognitively and emotionally—to dynamic temporal patterns produced by other humans in context of affiliation."

This combination of related adaptations was biologically motivated by co-occurrence of bipedalism, expanding brain size, and altricialization (Cross & Morley 2008; Dissanayake, 2008) and was fundamental to human survival. This is why, according to Dissanayake, proto-musical behavior produces such strong emotions, and activates brain areas involved in ancient mechanisms of reward and motivation, the same areas that are involved in satisfaction of most powerful instincts of hunger and sex.

A related theory of music origins is proposed by Parncutt (2008). He suggests that prenatal exposure to "the complex web of associations among patterns of sound, movement and emotion that characterize music" "creates a mother schema" that promotes postnatal survival. In this way, Parncutt suggests, one difficulty is overcome: the issues of music adaptivity and emotionality are dissociated, while both are supported. Many experiences of musical emotionality are explained, which otherwise seem mysterious. This might further be related to the origin of religion. Both, music and religion, he suggests, might be byproducts of prenatal experiences and the adaptive value of postnatal infant-mother bonding.

Mithen (2007) presents an impressive array of evidence that Neanderthals possibly have had proto-musical ability. He argues that music and language have evolved by differentiation of early proto-human voice sounds "Hmmmm" undifferentiated proto-music-language. The development was facilitated by vertical posture and walking, which required sophisticated sensorimotor control, a sense of rhythm, and possibly ability for dancing.

The differentiation of Hmmmm, he dates to after 50,000 BP. Further evolution toward music occurred for religious purposes, which he identifies with supernatural beings. Currently music is not needed, it has been replaced by language, it only exists as inertia, as a difficult to get rid off remnant of the primordial Hmmmm. An exception could be religious practice, where music is needed since we do not know how to communicate with gods. (I have a difficulty with dismissing Bach, Beethoven, or Shostakovich in this way; as well as with the implied characterization of religion, and discuss my doubts later).

Mithen explains why music is often perceived as a conversation, and why we feel it as having a meaning, both of these are remnants of Hmmmm. Onomatopoeia is also a survival of Hmmmm. Among a number of properties of music explained by Mithen, I would emphasize relation of music to emotions, this was present in original Hmmmm. Songs recombine language and music into original Hmmmm, however Mithen gives no fundamental reason or need for this recombination.

Mithen summarizes the state of knowledge about vocalization by apes and monkeys. Unlike older views, calls could be deliberate, however their emotional-behavioral meanings are probably not differentiated; this is why primates cannot use vocalization separately from emotional-behavioral situations (and therefore cannot develop language), this area is still poorly understood. While addressing language in details, Mithen (and other scientists as well) give no explanation for why human learn language by about age of five, but the corresponding mastery of cognition takes the rest of lifetime; steps toward explaining this are taken in Perlovsky (2006c,d; 2009a,b,d) and summarized in this review.

Mithen's view on religion contradicts the documented evidence for relatively late proliferation of supernatural beings in religious practice (Jaynes, 1976), and to mathematical and cognitive explanations for the role of religiously sublime in workings of the mind (Perlovsky, 2001; Levine & Perlovsky, 2008a).

Juslin and Västfjäll (2008) analyze mechanisms of musical emotions. They emphasize that in the multiplicity of reviews considering music and emotions, the very use of the word 'emotion' is not well defined. They discuss a number of neural mechanisms involved with emotions and different meanings implied for the word 'emotion'. I would mention here just two of these. First, consider the so called basic emotions, which are most often discussed; we have specific words for these emotions: fear, sexual-love, jealousy, thirst... Mechanisms of these emotions are related to satisfaction or dissatisfaction of basic instinctual bodily needs such as survival, procreation, a need for water balance in the body... An ability of music to express basic emotions unambiguously is a separate field of study. Second, consider the complex or 'musical' emotions (sometimes called 'continuous'), which we 'hear' in music and for which we do not necessarily have special words. Mechanisms and role of these emotions in the mind and cultural evolution are subjects of this review.

Levitin (2008) classified music in six different types, fulfilling six fundamental needs, and (as far as I understood him) eliciting six basic emotions. He suggests that music has originated from animal cries and it functions today essentially in the same way, communicating emotions. An ability to communicate emotions with voice and to correctly perceive emotions in voice has given and continues to give evolutionary advantage and is the basis for emotional intelligence. Emotions motivate us to act and neural connections facilitating this are bidirectional, action and movement may elicit emotions: "emotions and motivation are two sides of the same evolutionary coin." It is more difficult, he writes, "to fake sincerity in music than in spoken language." The reason that music evolved this way as an 'honest signal' because it "simply" co-evolved with brains "precisely to preserve this property." (Given the fact that even as simple animals as birds can fake their cries (Lorenz, 1981) I have my doubts about this "simply;" further doubts arise as soon as we think about actors, singers, and poets, not only contemporary professionals, but also those existing in traditional societies (Meyer, Palmer, & Mazo, 1998) since time immemorial.)

Mathematical modeling of the mechanisms of music perception and musical emotions was considered in (Purwins, Herrera, Grachten, Hazan, Marxer, & Serra, 2008a,b; Coutinho & Cangelosi, 2009). These modeling approaches can be used to obtain and verify predictions of various theories.

In the following sections we review mechanisms of music evolution from differentiation of original proto-music-language to its contemporary refined states. Discussions of mechanisms that evolved music from IDS to Bach and Beatles in previously proposed theories are lacking or unconvincing. Why do we need the virtual infinity of "musical emotions" that we hear in music (e.g. in classical Western music)? Is it an aberration or do they address potentially universal human needs? Dissanayake (2008) suggests that this path went through ceremonial ritualization, due to "a basic motivation to achieve some level of control over events..." If "for five or even ten centuries... music has been emancipated from its two-million year history and its adaptive roots says more about the recency and aberrance of modernity..." Cross & Morley (2008) argue against this conclusion: "...it would be impossible to remove music without removing many of the abilities of social cognition that are fundamental to being human." He concludes that "there are further facets to the evolutionary story (of the origins of music) requiring consideration. Investigation of the origins, emergence and nature of musical behaviors in humans is in its early stages, and has plenty more to reveal." In the following we review a novel hypothesis that clarifies some of these remaining "further facets" and provides bases for further research in several directions.

4.12.5 Differentiation and Synthesis

Here we expand on the previous discussion of differentiation and synthesis in section 4.7 and 4.10. The knowledge instinct operates in the dual hierarchy of the mind with two main mechanisms, differentiation and synthesis Perlovsky (2006d, 2007, 2008). At every level of the hierarchy it drives the mind to achieve detailed understanding by creating more specific, diverse and detailed concepts—this is the mechanism of differentiation. At the same time (as we discussed), KI drives us to understand various situations and abstract concepts as a unity of constituent notions. This mechanism of KI operating across hierarchical levels creates higher meanings and purposes—this is a mechanism of synthesis.

The main "tool" of differentiation is language. Language gives our mind a culturally evolved means to differentiate reality in great detail. The evolution of language required neural rewiring of circuits controlling vocalization. Vocal tract muscles in animals are controlled from an old emotional center and voluntary control over vocalization is limited (Deacon, 1989; Schulz, Varga, Jeffires, Ludlow & Braun, 2005; Davis, Zhang, Winkworth, & Bandler, 1996; Larson, 1991). Humans, in contrast, possess a remarkable degree of voluntary control over voice, which is necessary for language. In addition to the old mostly involuntary control over vocal tract human have conscious voluntary control originating in cortex.

Correspondingly, conceptual and emotional systems (understanding and evaluation) in animals are less differentiated than in humans. Sounds of animal cries engage the entire psyche, rather than concepts and emotions separately. A well-known example is differentiated calls of vervet monkeys (e.g. see a review in Seyfarth & Cheney, 2003). The calls convey information about different types of predators nearby; however understanding of a situation (concept of danger), evaluation (emotion of fear), and behavior (cry and jump on a tree) are not differentiated, each call is a part of a single concept-emotion-behavior-vocalization psychic state with very little differentiated voluntary control (if any).

Emotions-evaluations in humans have separated from concepts-representations and from behavior (For example, when sitting around the table and discussing snakes, humans do not jump on the table uncontrollably in fear, every time "snakes" are mentioned). We hypothesize that gradual differentiation of psychic states with a significant degree of voluntary control over each part gradually evolved along with language and the brain rewiring.

Therefore, language contributed not only to differentiation of conceptual ability, but also to differentiation of psychic functions of concepts, emotions, and behavior. This differentiation destroyed the primordial synthesis of psyche. With the evolution of language human psyche started losing its synthesis, wholeness. Whereas for animals every piece of "conceptual knowledge" is inextricably connected to emotional evaluation of a situation, and to appropriate behavior, satisfying instinctual needs, this is not so for humans. Most of the knowledge existing in culture and expressed in language is not connected emotionally to human instinctual needs. This is tremendously advantageous for development of conceptual culture, for science, and technology. Humans can engage in deliberate conversations, and if disagree, do not have to come to blows. But there is a heavy price that humans pay for this freedom of conceptual thinking: human psyche is not automatically whole. Human knowledge accumulated in language is not automatically connected to instinctual needs; sometimes culturally developed conceptual knowledge contradicts instinctual needs inherited from the animal past. Moreover, various parts of knowledge may contradict each other. As discussed, synthesis, the feeling of being whole is closely related to successful functioning of the highest models at the top of the hierarchy of the mind, which are perceived as the meaning and purpose of life. Therefore contradictions in the system of knowledge, the disconnects between knowledge and instincts, the lost synthesis, lead to the internal crises and may cause clinical depressions. When psychic states missing synthesis preoccupy the majority of a population, knowledge loses its value, including knowledge and value of social organization and cultural calamities occur, wars and destructions (Perlovsky 2006b,e,f, 2007, 2008; Diamond 1997). The evolution of culture requires a balance between differentiation and synthesis. Differentiation is the very essence of cultural evolution. But it may lead to emotional disconnect between conceptual knowledge and instinctual needs, to the lost feeling of the meaning and purpose, including the purpose of any cultural knowledge, and to cultural destruction. Theoretical and experimental evidence suggest that different languages maintain different balances between the emotional and conceptual (Guttfreund 1990; Balaskó & Cabanac 1998; Buchanan, Lutz, Mirzazade, Specht, Shah, Zilles, et al, 2000; Harris, Ayçiçegi, & Gleason 2003; Perlovsky 2007).

4.12.6 Differentiated Knowledge Instinct and Musical Emotions

Here we discuss the main hypothesis of this section: what constitutes the fundamental role of musical emotions in evolution of consciousness, cognition, and culture.

As discussed, the balance between differentiation and synthesis is crucial for the development of cultures and for emergence of contemporary consciousness. Those of our ancestors, who could develop differentiated consciousness, could better understand the surrounding world, and better plan their life had evolutionary advantage, if in addition to differentiation they were able to maintain the unity of self required for concentrating will. Maintaining balance between differentiation and synthesis gave our ancestors evolutionary advantage. Here we examine the mechanisms by which music helps maintaining this balance. The main hypothesis of this paper is that maintaining this balance is the very fundamental role that music plays and the reason for evolution of this otherwise unexplainable ability.

History keeps a long record of advanced civilizations, whose synthesis and ability to concentrate its will was undermined by differentiation. They were destroyed by less developed civilizations (barbarians) who's differentiation lagged behind, but who's synthesis and will was strong enough to overcome great powers of their times. These examples include Akkadians overrunning Sumerians some 3 millennia BCE, barbarians overcoming Romans, and countless civilizations before and after these events. But I would like to concentrate on less prominent and more important events of everyday individual human survival from our ancestors to our contemporaries. If differentiation undermines synthesis, undermines the purpose and the will to survive, then differentiated consciousness and culture would never emerge.

Let us repeat, differentiation is the very essence of cultural evolution, but it threatens synthesis and may destroy the entire purpose of culture, and the culture itself (Perlovsky 2005, 2006b,e,f, 2007, 2009b). This instability is entirely human, it does not threaten the animal kingdom because the pace of evolution and differentiation of knowledge from ameba to primates was very slow, and instinctual mechanisms of synthesis apparently evolved along with the brain capacity. This situation drastically changed with the origin of language; accumulation of differentiated knowledge vastly exceeded biological evolutionary capacity to maintain synthesis. Along with the origin of language another uniquely human ability evolved, the ability for music. Based on the previous discussions in this book, we propose a scientific hypothesis that music evolved for maintaining the balance between differentiation and synthesis. After reviewing arguments, we discuss empirical and experimental means by which this hypothesis can be verified.

Many scientists studying evolution of language came to a conclusion that originally language and music were one (Darwin, 1871; Cross, 2008a; Masataka, 2008). In this original state the fused language-music did not threaten synthesis. Not unlike animal vocalizations, sounds of voice directly affected ancient emotional centers, connected semantic contents of vocalizations to instinctual needs, and to behavior. This way Jaynes (1976) explained stability of great kingdoms of Mesopotamia up to 4,000 years ago. This synthesis was a direct inheritance from animal voicing mechanisms, and to this very day voice affects us emotionally directly through ancient emotional brain centers (Panksepp & Bernatzky, 2002; Trainor, 2008).

We would like to emphasize the already discussed fact that since its origin language evolved in the direction of enhancing conceptual differentiation ability by separating it from ancient emotional and instinctual influences (here we mean "bodily" instincts, not instincts for knowledge and language). While language was evolving in this more conceptual and less emotional direction, we suggest that 'another part' of human vocalization evolved toward less semantic and more emotional direction by enhancing already existing mechanisms of voice-emotion-instinct connection. As language was enhancing differentiation and destroyed the primordial unity of psyche, music was reconnecting differentiated psyche, restoring the meaning and purpose of knowledge and making cultural evolution possible. Was this process equally successful in every culture? Probably not, but this is a separate field of study for future research.

This was the origin and evolutionary direction of music. Its fundamental role in cultural evolution was maintaining synthesis in the face of increasing differentiation due to language. We now return to the basic mechanisms of the mind, including KI and analyze them in more details in view of this hypothesis.

Discussing KI in previous sections we described the mathematical model of its mechanism, a mental "sensor" measuring similarity between concept-models and the world and related mechanisms of maximizing this similarity. But clearly it is a great simplification. It is not sufficient for the human mind to maximize an average value of the similarity between all concept-models and all experiences. Adequate functioning requires constant resolution of contradictions between multiple mutually contradicting concepts and between individual concepts quickly created in culture and slowly evolving primordial animal instincts. Human psyche is not as harmonious as psyche of animals. Humans are contradictory beings; as Nietzsche (1995/1876) put it, "human is a dissonance." Those of our ancestors who were able to acquire differentiated contradictory knowledge and still maintain wholeness of psyche necessary for concentration of will and purposeful actions had tremendous advantage for survival.

Therefore, we suggest that KI itself became differentiated. It was directed not only at maximizing the overall harmony, but also at reconciling constantly evolving contradictions. This is a hypothesis that requires theoretical elaboration and experimental confirmation. As discussed, emotions related to KI are aesthetic emotions subjectively felt as harmony or disharmony. These emotions had to be differentiated along with KI. Consider high value concepts such as one's family, religion, or political preferences. These concepts 'color' with emotional values many other concepts; and every contradictory conceptual relation requires a different emotion for reconciliation, a different dimension of an emotional space. In other words, a high value concept attaches aesthetic emotions to other concepts. In this way each concept acts as a separate part of KI: evaluates other concepts for their mutual consistency; this explains the notion of the differentiated knowledge instinct. Virtually every combination of concepts has some degree of contradictions. Psychologists sometimes For example The number of combinations is practically infinite (Perlovsky, 2006d). Therefore aesthetic emotions that reconcile these contradictions are not just several feelings for which we can assign specific words. There is an uncountable infinity, continuum of aesthetic emotions, and most likely the dimensionality of this continuum is huge. We feel this continuum of emotions (not just many separate emotions) when listening to music. We feel this continuum in Palestrina, Bach, Beethoven, Mozart, Chaikovsky, Shostakovich, Beatles, and Eminem… (and certainly this mechanism is not limited to western cultures).

I would mention that Spinoza (2005/1677) was the first philosopher to discuss the multiplicity of emotions related to knowledge. Each emotion, he wrote, is different depending on which object it is applied to. There is a principled difference between multiplicity of aesthetic emotions and 'lower' emotions corresponding to bodily instincts. Those emotions, as discussed, are referred to as 'basic' emotions in psychological literature (e.g. see Juslin & Sloboda 2001; Sloboda & Juslin 2001; Juslin & Västfjäll, 2008). As discussed, psychologists identify them; they all have special words, such as 'rage' or 'sadness.' Levitin (2008) suggests that there are just six basic types of songs, basic emotions related to basic instinctual needs. But Huron (1999) has already argued that this use of music for basic needs is just that, a utilitarian use of music, which evolved for a much more important purpose that cognitive musicologists had not yet been able to identify. Sloboda & Juslin (2001) emphasized that musical emotions are different from other emotions. Emotions related to "mismatch" and "discrepancies" were discussed in (Frijda, 1986; Juslin & Sloboda, 2001). It is proposed here that musical emotions have evolved for synthesis of differentiated consciousness, for reconciling contradictions that every step toward differentiation entails, for creating a unity of differentiated Self.

The referenced literature suggests that music has two interrelated purposes fundamental to the functioning of individual minds and to evolution of the mind and culture. The first purpose is to differentiate aesthetic emotions. Music creates differentiated emotions required to reconcile conceptual contradictions. The second purpose is to connect concepts to instinctual needs (including KI). Whereas language separates conceptual knowledge from instincts and emotions, music reconnects these ties. Both musical functions suggested here are scientific hypotheses that should be and are going to be further explored theoretically and verified experimentally.

4.12.7 Empirical Evidence and Tests

The previous section reviewed the hypothesis about the fundamental role and function of musical emotions in evolution. Here we review empirical evidence for this hypothesis. First, we consider historical evidence for parallel evolution of culture, consciousness, and musical styles. Much evidence has been accumulated concerning the latest 3000 years of cultural evolution, over which recorded evidence exists (Weiss and Taruskin 1984; Jaynes 1976; Perlovsky 2005, 2006b,e, 2008). This evidence demonstrates that advances in consciousness and cultures were paralleled by advances in differentiation of musical emotions. Here we select few examples from this history. Second, we consider future directions for laboratory psychological and neuroimaging experiments that could verify this hypothesis and experimentally connect differentiation of musical emotions to synthesis of consciousness. Several groups of psychologists plan these experiments.

4.12.7.1 Role of Music in Cultural Evolution (from King David to the 20th Century)

Before getting to empirical examples we recollect the main theoretical ideas. Interaction of differentiation and synthesis considered in the previous sections is a general law of KI operations, characteristical of any epoch in human history. Accelerated differentiation of everyday life tips the balance of the everyday and the highest. "It is difficult to keep the scissor blades together." (Brodsky 1991/2000/2003). It is difficult because the condition of the creative process is the combination of oppositions, differentiation and synthesis. Their complex dynamics determines the development of culture. When unity within the soul is achieved (synthesis), creative energy is directed at exploration of the outer and inner world, at widening the sphere of conscious – that is, diversification-differentiation of everyday concepts and emotions. (So, Judeo-Christian synthesis prepared the ground for understanding that human is the source of creative spirit, and this formed conditions for emergence of scientific thinking, although it took thousands of years to come to fruition. Only in the 17th c. Descartes completed "expelling spirit from matter" and Newton, following him, could think about completely causal, that is scientific, explanation of the material world).

In the process of history, diversity of everyday life gets complicated and overtakes concepts of the highest, which have served as a foundation for inspiration-synthesis. Lagging synthesis leads to a discord in the soul – concepts of the highest purpose do not correspond to everyday way of life, to variety of concepts and emotions, leading to a decline of culture. (So scientific thinking destroys ancient religious synthesis). Overcoming crises and continuing the cultural process demands new concepts of the highest purpose, new synthesis, corresponding to a new level of the differentiation of psyche.

With increasing differentiation, synthesis requires ever increasing efforts of an individual human being. Balancing these two aspects of consciousness is difficult and is achieved through understanding of the purpose of life; Jung (1921) called this the highest aim of every human life. Similar was Schopenhauer's idea of individuation (1819). Even more radical was Kant (1790), who wrote that consciousness of the purposiveness coincides with the Christian ideal of sainthood. Consciousness and culture are developed on the edge of differentiation and synthesis. Too strong a synthesis fuses the conscious and the unconscious together into a fuzzy undividedness, the need and ability for the new disappears, as in pre-historic consciousness. Prevalence of synthesis is characteristic of Eastern cultures, striving for the peace of the soul. A payoff for the peace of soul is millennia of cultural immobility. Prevalence of differentiation is characteristic of Western cultures, when differentiation overtakes synthesis, the meaning of life disappears, and creative potential is lost in senselessness.

What has been the role of music in this complex process of "keeping the scissor blades together"? Let us start on the promised short historic excursion. Jaynes (1976) analyzed the evolution of consciousness during the last 11,000 years. Weiss and Taruskin (1984) analyzed the evolution of musical styles using available data during the last 3,000 years. (throughout this section we refer to this publication as W&T, and sometimes we use quoted statements without refs.).

These two sets of changes in consciousness and in music were aligned in (Perlovsky 2005, 2006b,e, 2008; also Jaynes 1976 analysis was extended by adding the idea of synthesis). This alignment demonstrated first, that during the states of strong synthesis, advances in consciousness were driven by differentiation and music differentiated "lower" emotions, and second, that differentiation violated synthesis. To restore synthesis, music differentiated emotions of the "highest". These emotions helped to understand the violation of synthesis by bringing it from the unconscious into consciousness. The conscious understanding helped to cope with the violated synthesis and to continue the process of conceptual differentiation of consciousness and cultural evolution. From this continuous millennial process here we select several examples illustrating that every step in conceptual differentiation was paralleled by powerful advances in music, first, bringing a new level of emotional differentiation to everyday life, and second emotional differentiation of the "highest", which helped to restore synthesis.

Contemporary Western music originated from church and sinagogal singing; according to W&T "Psalmody (the singing of psalms) is surely the oldest continuous musical tradition in Western civilization." However, the first Biblical description referring to King David time (3,000 years ago) refers to "the clangorous noise of instruments... reminds the modern reader of no Western form of divine service... (similarly does a scene) of David dancing before the arc of God." Why? Possibly because there were no irresolvable contradictions in the souls of David and his contemporaries, the monotheistic idea was a sufficient basis for synthesis. Human imperfections were sins, for which one had to be accountable before God, but the notions of sin, freedom, and personal responsibility were not yet sufficiently differentiated to precipitate existential crises. This relatively undifferentiated type of consciousness we see in the book of prophet Amos written in the 8th c. BCE, 250 years after David. Consciousness presented in this book, was characterized as follows: "In Amos there are no words for the mind or think or feel or understand or anything similar whatsoever; Amos never ponders anything in his heart. In the few times he refers to himself, he is abrupt and informative…" (Jaynes 1976); his speech, voice, words, emotional and conceptual contents were fused, there were no deliberation, no arguments, no choices to be made. In this period of fuzzy consciousness, music of the divine service, like all creative forces, was directed at differentiation.

However, a new type of consciousness was already rising; consciousness with self-reflection and internal contradictions. Although the prophecy of Isaiah took place only one generation after Amos, Isaiah's consciousness was ahead of his contemporaries. The impending catastrophe that he foresaw created tensions in his soul between conscious and unconscious. This tension appeared in his vision as an antiphony of the voices of Seraphims. For the first time the principle of antiphony was mentioned in the Bible, the split choirs answering back and forth, which was to become a foundation of psalmody in Jewish and Christian divine service: "Seraphim... one cried to another, and said, Holy, holy, holy is the Lord of hosts." (Is. 6, 1-4.) "The words sung by the Seraphim entered the Jewish liturgy... and were later adopted by the Christian church..."

Development of consciousness in Ancient Greece, Israel, and China remarkably coincide. In the 6th c. BCE the first Greek philosopher Thales repudiated myths, demanded conscious thinking, and pronounced the famous "know thyself". In Israel, Prophet Zechariah (Zech. 3-4) forbade prophecy, an outdated and already dangerous form of thinking; he demanded conscious thinking. Confucius (5th c. BCE) wrote "when we see men of a contrary character, we should turn inwards and examine ourselves", and his contemporary Lao-tzu, "it is wisdom to know others; it is enlightenment to know one's self." Conscious thinking created a discord between personal and unconscious-universal, led to a feeling of separateness from the world; tensions appeared in psyche, which were mirrored in antiphonal singing. – Forms of music appeared, corresponding to the forms of consciousness. – Singing of split choirs symbolized differentiated nature of the highest principles, and brought closer to consciousness the feel of the split in psyche. Antiphonal singing, appealing to conscious and to unconscious, drew them closer, linked the feeling of the split with conscious perception of "self-world" relationships, and restored synthesis. Antiphon as a generally accepted form of divine service is mentioned in the Bible for the first time in the book of Nehemiah (Neh. 12, 27-43) in 445 BCE, just a century after Zechariah and Thales' "know thyself."

Let us move forward by two millennia, to the Renaissance (the 13th -16th c.). In the beginning of the Renaissance (the 13th -14th c.), synthesis was strong, backed up by both, a new symbol of the greatness of human reason and by ancient religious mystical symbols; the result was a creative explosion. From the "objective," music moved toward differentiating everyday human feelings. In the 14th c. the first musical avant-garde emerged; Ars Nova (The New Art) used notes of variable durations for further differentiation of emotions. Pope John XXII (1323) criticized the new music: "By… dividing of beats… the music of the Divine Office is disturbed with these notes of quick duration. Moreover, they hinder the melody with hockets (interruptions), they deprave it with discants (high-voice ornamental melodies), and… pad out the music with upper parts made out of secular songs… The voices incessantly rock to and fro, intoxicating rather than soothing… devotion… is neglected, and wantonness… increases." The Pope as if foresaw the crisis of culture due to the lost beliefs and synthesis. The Christian symbol was losing autonomous power in the human soul. Trouble was all over Europe, strife among nations and social classes, Papal exile, schism within the Church, The Great Plague. The catastrophe coincided with a lost unity within the soul, and again, chasing the lost wholeness, a cycle of the restoration of synthesis wound up.

What kind of music could inspire people, when the power of the mysterious was lost and the dominating idea was humanism, the power of human reason? Beginning in the Renaissance, a musical system of tonality was developed for differentiation of emotions, and for connecting the everyday with sublime.

Music connecting differentiated emotions with the sublime emerged in the 15th c. John Dunstable, according to contemporary witnesses, changed all "music high and music low," music became more consonant and euphonious. Melody and rhythm were concentrated in the top part, supported by chordal harmonies.

"Harmonies exalted even heaven... like angelic and divine melodies... (As if the) songs of angels and of divine Paradise had been sent forth from the heavens to whisper in our ears an unbelievable celestial sweetness." (W&T, 81-82).

The Renaissance synthesis was based on humanism, human values: "music's true purpose and content... (is its) power to move" emotions (Glareanus, the 16th c.). This thought "a medieval thinker would have found incomprehensible... The new Renaissance attitude... valued the natural, spontaneous gift of the artist over the application of reason and mastery of theoretical doctrine." Attitude to emotions in music was changing. After 3500 years of monotheism man was becoming (to some extent) a master of the self. Untamed emotions were no longer considered a morbid threat to society, self, and spiritual interests. Humanistic ideal had inspired the Renaissance man to look for increasingly stronger emotions – and this search continues today (albeit not without interruption).

The highest ideal of Christianity, improvement of inner spiritual life traditionally demanded repudiation of the material world perceived as temptation and distraction from the highest spiritual purpose. The best way to achieve the ideal of sainthood was supposed to be a monastic way of life and rejection of secular life. However, rejection of the world acknowledged the absolute power of evil projected in the material world. By the 15th c. the ascetic ideal came to contradictions with developing rational thinking and the emerging capitalistic economy. Reformation in the 16th c. accepted that the highest human calling was in perfecting the inner spiritual world as well as the outer material world (and material conditions of one's life). The religious ideal was reconciled with new consciousness.

The Reformation reduced the absoluteness of the split between spiritual and material, good and evil, - the contradiction between good and evil was taken from the heights of Heaven and the depths of Hades and placed into the human soul. Consequences were on one hand an inconceivable acceleration of the development of capitalism and improvement of material conditions of life. On the other, the autonomy of religious symbols was lost; their unconscious contents were to a large extent transferred into consciousness. The fundamental contradiction of human nature between finite matter and infinite spirit, which formed the mystical foundation of Christianity, was brought by the Reformation into everyday culture and made a part of collective consciousness. Tragic tensions originally projected onto the Christian symbol were assimilated by human psyche. Tensions in the human soul reached the maximum.

Luther (1538) saw in music the synthetic power that unifies the Word of God with human passions: "Therefore... message and music join to move the listener's soul... The gift of language combined with the gift of song was only given to man... (so that he proclaims) the word through music."

Naive humanism of the 15th c. barely glimpsed into the contradictions of human thoughts. Consciousness of the mind's internal contradictions was an achievement of the Reformation, and this consciousness required new forms of synthesis to restore wholeness. In search of synthetic forms of art creative minds turned to the epoch of crisis long gone, when salvation was found in art. From the books of Plato, Aristotle, and other authors of antiquity it was known that tragic musical drama in Ancient Greece created catharsis, an intimate bond with the human soul, which miraculously calmed discontent, soothed character and behavior. 'Radical humanists' in the sixteenth century sought to recover the true music of antiquity, which according to their ideas was in close connection with rhetoric, the art of orators and actors. A literary expression of these ideas was given by Vincento Galilei (Florence, 1588; see W&T). A new form of music, 'musical speech', or recitative, quickly led to a true opera "Orpheus" by Claudio Monteverdi (1600) and made a profound influence on the following development of Baroque music.

The Baroque was full of dualism and drama, expressing tensions imposed by the Reformation. It is a world searching for differentiated synthesis. Dualism was embodied in a new musical style, where opposition was emphasized: Vocal against instrumental, solo against ensemble, melody against bass, dynamic levels were contrasted, opposition of the dominant and tonic, all expressed emotional tension and resolution. The role of dissonances increased, and modulations became commonplace expressing more and more complex emotions in their continuous flow. Creating emotions was becoming the primary aim of music; composers strived to imitate speech, the embodiment of the passions of the soul. At the same time conceptual content of texts increased, "the words (are to be) the mistress of the harmony and not its servant," wrote Monteverdi. This became the main slogan of the new epoch of Baroque music. Thus, conscious aims of Baroque music were differentiation of emotions in parallel with synthesis of conceptual and emotional.

Tensions in human sole created by Reformation continued to propel a search for higher and higher synthesis, requiring stronger differentiation of emotions of the highest ideal, corresponding to the consciousness of the split between finiteness of human material being and infiniteness of spiritual aspirations. Until the end of the 16th c. dissonances were used sparingly, for a short pause, and mainly in secular music. Beginning in the seventeenth century dissonances were used more often, emphasizing the dramatic effect. A dissonance was always followed by a resolution in a consonant chord, later several dissonant chords were used in a row, increasing tension. The heightened sense of drama in musical dissonances corresponded to the tension between conceptual and emotional, material and spiritual, which in the result of the Reformation where assimilated by human heart and soul. Music became extremely expressive, conveyed passionate human emotions; theory of major and minor scales were developed for this purpose, chromatic scale was used. Chorale was unified with counterpoint, harmony with polyphony. These new musical forms were perfected in works of Buxtehude and then Bach.

The most complex and sublime form of polyphonic music was acquired in fugue. Fugue is a conversation of several musical voices, in which a topic "flies" from one voice to another; voices could talk politely or argue, interrupting each other. In Bach's fugues a human arguing with oneself turns to God or to the highest in oneself. Whereas old psalms affirmed an existence of the objectively sublime, as some collective purpose far removed from individual experiences, fugue expressed emotions of ones own contradictions in quest for the highest. Fugue was a way of individual consciousness turned to sublime, a combination of differentiation and synthesis. Rational understanding of Church service introduced by the Reformation interacted in music with the highest spiritual values and mystical feelings of sublime, created during thousands of years by monotheistic religions.

However, the Reformation has laid unbearable responsibility on an individual and created too much tension within the human soul - humankind is not ready yet for individual consciousness. The string of tension connecting conscious and unconscious broke. Rational consciousness that came after the Baroque rejected mystery of sublime differentiated in fugue. Music that was natural to Bach seemed too intellectual and "not natural" to the next generation.

Differentiation of consciousness and development of corresponding musical forms tremendously accelerated. To fit the content of our discussion within the limits of this review, however, I would have to skip through fascinating developments of Rococo, Classicism, and Romanticism and to move to few examples of the 20th c. In the 20th c. all areas of human spiritual endeavor became more entangled than ever before. Attempts to create formalized mathematical logic in the 2nd half of the 19th c. were soon repeated in the idea of dodecaphonic music developed by Schoenberg. As if foreseeing the horrors of the coming world wars, Schoenberg aspired to move beyond emotions that could be created by tonal music. Much of music of the 20th century, for example those of thriller movies, evolved from Schoenberg's idea. This music often cannot be even written in traditional musical notations. In all areas of art "modern" looked for differentiation of human unconscious. The opposite tendency to restoring synthesis at all cost began at the same time, but only later, in the 1970s was recognized as such and called "postmodern."

Differentiation and synthesis evolved in parallel often intersecting in lives of individual artists. The contradiction can be seen in the art of Schoenberg. He formulated an atonal idea (dodecaphony) as a formal rule, but attempted to express in music unverbalizable nature of God. For more than sixteen years he worked on "Jacob's Ladder" and "Moses and Aron," still both works remained unfinished. The formal dodecaphonic rule did not fit the needs of human soul. I'd mention that similar was the fate of mathematical formalism, which inspired Schoenberg; in the 1930s Gödel proved its inconsistency.

The very idea of "objective" formal art contained antinomy manifested in the most unexpected ways. Malevich declared the aim of Suprematism – to free art from any symbolic content – but his "Black Square" was interpreted as a symbol of impenetrable unconscious content. In "Ulysses" Joyce created a form of language to express a 'stream of consciousness', but an almost complete absence of consciousness was the outcome. C. Jung uses "Ulysses" to characterize a significant part the 20th c. art and collective consciousness as follows: "...A passive, merely perceiving consciousness, a mere eye, ear, nose, and mouth, a sensory nerve exposed without choice or check to... a stream of physical happenings... The stream... not only begins and ends in nothingness, it consists in nothingness. It is all infernally nugatory... Today it still bores me as it did then (in 1922). Then, why do I write about it?... (It) is a collective manifestation of our time... the collective unconscious of the modern culture... the modern artist immerses into destructive processes, to affirm in destructiveness the unity of his artistic personality... We still belong to the Middle Ages... For that alone would explain... why there should be books or works of art... (like) "Ulysses." They are drastic purgatives... for the soul... which is of use only where the hardest and toughest material must be dealt with." (Jung, 1934). Those agreeing with Jung about roots of Joyce popularity would find many similar examples in music.

In music, like in visual art and philosophy, two contrary historical tendencies of evolution of consciousness collided again, differentiation and synthesis. It's not surprising that changes in musical forms paralleled visual arts, philosophy, and science. (Differentiation of self, as a penetration into the depths of unconscious was manifest in the psychology of Freud, paintings of Pollock, music of Scriabin and Shostakovich, to name just few). Differentiation, however destroyed the wholeness of the world perception, and a contrary tendency emerged, postmodern, as a striving for synthesis based on the simplest notions (such as music of Cage). Whereas in the past centuries differentiation may have dominated one epoch and synthesis another, in the 20th c. all mixed up. While Modernism sought depths of self, Postmodern with equal force rushed to simplicity of the bases of aesthetic. The opposing tendencies of differentiation and synthesis were present in conscious and unconscious of an individual composer.

Mass culture is a logical step in evolution of consciousness, in interaction of differentiation and synthesis. There is a chasm between differentiated concepts existing in culture and capacity of a single person to assimilate this culture, while preserving synthesis within one's soul. Is this chasm unprecedented and unique for our times? Was this chasm smaller for Aristotle and Ancient Greek crowds? Surprising animalistic and satanistic styles of some rockers and rappers could be understood if we compare them to Ancient Greek dithyrambs. The dithyramb was an ancient way of creating synthesis, connecting the sublime with bestial unconscious bases of psyche. The rift between conscious and unconscious threatens the death of culture and "demands restoratory sacrifices" (Nietzsche 1876). Rap (hip-hop) is contemporary dithyramb very similar in musical and performance style, restoring the connection between conscious and unconscious. In both dithyramb and rap – quite regular thoughts are cried out at the edge of frenzy. As in Ancient Greece 2,500 years ago, so today in a complex multiform culture, people, especially young people, are losing their bearings. Words no longer call forth emotional reactions, their prime emotional meaning is lost. By shouting words along with primitive melody and rhythm, a human being limits his or her conscious world, but restores synthesis, connection of conscious and unconscious. An internal world comes to wholeness, reunites with a part of the surrounding culture.

As postmodern art and music in particular was a return to pre-Aeschylean, Apollonian consciousness of pure notions – so Rap is a natural continuation of postmodern: Dionysian breaks forth into Apollonian consciousness. These types of consciousness antiquated about 2,500 years ago. But consciousness does not whirl in a closed circle. Conceptual and emotional contents of contemporary culture have become much richer, and the previously unseen poles of differentiation are to be unified by the coming synthesis.

Leaning upon scientific analysis of the mind functioning in previous sections, this section reviewed changes of forms of consciousness parallel to changes of musical forms. Summarizing, I would emphasize that music is the most mysterious ability of a human soul; it contains differentiating and synthesizing powers. Necessity governs relationships between these powers: when rocking toward differentiation concepts lose their meanings and culture is destroyed, but when rocking toward synthesis strong emotions nail down thoughts to traditional values. Both lead to a slowdown of cultural evolution. As no other art, music can forestall cultural slowdown. Music transports reality into the hearts of listeners and restores a possibility of continuation of culture. But will the unity of differentiation and synthesis prevail over life? Or will our entire culture be torn into shreds? Answering this question is one of the directions of further development of ideas in this review as well as an empirical way of investigating scientific validity of the reviewed theory of musical emotions.

4.12.7.2 Future Laboratory Experiments

Laboratory experimentation should be directed at operational definition and measurements of musical emotions. According to the reviewed hypothesis, the function of musical emotions in cognition is to restore synthesis, when it is damaged by differentiation. Such a condition is similar to cognitive dissonance in psychology (Festinger, 1957). Therefore, well developed experimental techniques used to study cognitive dissonance can be used to study the proposed role of musical emotions in reconciling contradictions in consciousness. This approach would directly verify the review's proposal that the multiplicity of musical emotions is related to contradictions in consciousness among conceptual knowledge. First, various types of cognitive dissonances can be created in subjects using standard techniques (Akerlof & Dickens 2005). Second, various types of music can be assessed for their efficiency in reconciling specific types of conceptual dissonances. In this way various types of music, which are known to create musical emotions (Juslin & Sloboda 2001; Steinbeis, Koelsch, & Sloboda 2006; Patel 2008) can be connected to reconciling specific dissonances (we would expect that the results would depend on psychological types of listeners, people who's feelings are less differentiated might be more affected by tonal music, whereas people consciously differentiating many emotions might be more susceptible to atonal music – this comment, however, is secondary to the main ideas discussed here); neuroimaging techniques can be used in parallel to identify the brain regions involved.

When significant experimental data are accumulated, dimensionality and structure of musical emotion space can be investigated by mathematical methods. Existing mathematical techniques of multidimensional scaling can be used. A future direction would be to develop methods for estimating dimensionality of a space of a very large dimension from limited number of measurements. Another direction for future mathematical research would be to develop techniques for exploring the notion of "continuous" musical emotions, as they are called in psychology. Of course, any number of measurements could only yield a finite number of data points. Can mathematical methods be developed to estimate a density of the space of musical emotions and a measure of their "continuum"?

Using existing and future mathematical techniques we would be able to explore the complexity of emotional spaces (structure, dimensions) and compare various types of music. It would be interesting to compare emotional spaces of Eminem and Beethoven to confirm or disprove various expectations.

The role of timbre in music and language might be related to the discussion in this review. Levitin (2006) writes that timbre characterizes individual performers more than any other aspect of music. Patel (2008) suggests that language uses timbre systematically more than music does. Is timbre evolved as "semantic," whereas melody "emotional"? Is harmony related to the mind hierarchy? Are these intuitions just shallow metaphors or meaningful, experimentally testable hypotheses related to the initial separation of voice into language and music, and to further evolution of cultures and consciousness?

We would like to emphasize possible directions for experimental verifications of the suggested mechanisms of KI and dual models, and their role in the mind functioning. The dual model of the neural mechanism connecting language and cognition can be studied using various neuroimaging techniques. A recent publication seems to support the dual model hypothesis (Franklin et al 2008). They have demonstrated that certain cognitions based in the right brain hemisphere in prelinguistic infants are rewired to the left hemisphere as language is acquired. Varying brain imaging techniques can be used to study more diverse connections between language brain areas and conceptual representation areas. Identifying brain modules and neural connections involved in the dual models and knowledge instinct was discussed in (Levine & Perlovsky, 2008a;b). Perlovsky, Bonniot-Cabanac, & Cabanac (2009) initiated psychological studies of the knowledge instinct.

4.12.8 Summary and Further Directions

Musical power over human soul and body has remained mysterious from Aristotle to the 20th century cognitive science. Contemporary evolutionary psychologists have recognized music as a cultural universal of tremendous power; still its fundamental role and function in cognition, its role in evolution of consciousness and culture have remained hidden. Here we reviewed historical and contemporary scientific hypotheses of the role and function of music, and concentrated on one hypothesis. It explains musical emotional mechanisms by relating them to primordial connections between voicing and emotions. It explains the role and function of music in differentiating emotions for the purpose of restoring the unity of self. Musical emotions help maintain a sense of purpose of ones life in face of multiplicity of contradictory knowledge, or what is called the "synthesis of differentiated consciousness."

According to this hypothesis, the origins of music are tied to the origins of language. Language emerged by differentiating the original unity of primordial self. Original psychic states of unified concept-emotion-behavior-vocalization were differentiated, so that concepts shed off their inextricable connections to emotions and motivation, and deliberate thinking-conversations became possible. Language was emerging. The price for this differentiation was the loss of the unity of self, lost concentration of will. Our ancestors, who could maintain concentration of will, while differentiating the knowledge about the world, received evolutionary advantage. Therefore the emotional part of primordial vocalization evolved into music.

As language and culture were evolving into a powerful system with tremendous differentiation of knowledge about the world and self, the number of contradictions grew combinatorially. Every combination of conceptual pieces of knowledge led to its own shades of contradictions. Therefore, maintaining motivation for this diversified knowledge required virtually infinite number of shades of motivations. Musical emotions are called "continuous" by psychologists of emotions. All of these emotions-motivations are related to knowledge, and therefore, since Kant, are called aesthetic emotions.

Can this hypothesis be verified by scientific empirical methods? One direction discussed in section 10.1 is to relate the changes in musical styles to the changes in cultures and consciousness. This connects evolution of music, consciousness, and cultures. A step in this direction was made in Perlovsky (2006b,e, 2008). It was suggested for example that antiphonal music appeared about 2500 years ago along with contemporary consciousness, when fundamental contradictions in human psyche started penetrating into consciousness and created psychic tensions. Tonality was developed beginning in the Renaissance, when instinctual and emotional human nature was consciously accepted, creating tensions in psyche with received ideas of spiritually 'high.' Buxtehude and Bach were developing music that could reconcile new contradictions brought in consciousness by the Reformation. Popular songs restore synthesis by connecting conceptual contents of lyrics with emotional contents of music. And contemporary rap music was suggested to have a similar style and function to Ancient Greek dithyrambs, namely to reconcile instinctual needs with (at least some) basic concepts in culture and language. Further researches in this direction are virtually unlimited. They should extend from details in global changes of consciousness and cultures to changes in lives of individual composers. In this regard it is interesting to mention what musicologists call the "swan song" phenomenon (Simonton 1997). Many composers created their most profound musical compositions in later years of their lives. Is it because synthesis becomes psychologically more important in later years? Examples of musical evolution in this review address western tradition, they should be extended to other cultures. Especial challenge is presented by tonal languages (e.g. Mandarin), in which melody might play both conceptual and emotional role. Is this conflation an advantage or an impediment for long-term cultural development?

Section 10.2 discussed laboratory experimental studies related to this hypothesis. Most of these are still future research. Among this, the first conceptual step would be to operationally define and model musical emotions. This has been related to the well developed methods of studying cognitive dissonance, an unpleasant feeling experienced when becoming conscious of contradictions in ones system of knowledge and beliefs (in other words, threat to synthesis due to differentiation). Cognitive dissonance has been an important psychological tool in developing Tversky and Kahneman (1974) theory of human irrationality (2002 Nobel Prize).

Laboratory experimental tests should be used to study emotional (melodic) contents of various languages vs. emotional contents of music developed in various cultures. These studies are difficult due to received prejudices. For example, the difference in emotionality between English and Italian people is often explained by climate, etc. But these explanations do not fit high emotionality of Russians. More fundamental studies are needed: which part of emotionality is related to behavior, cognition, and which to language alone? Does the increase in popularity of songs in English-speaking cultures compensate the reduced melodic contents of English language? Mathematical methods should be developed to study the spaces of musical emotions and ways to estimate large dimensional spaces and their structures from finite amount of measurements.

Classical psychological tests as well as brain imaging should be used to test the dual model, the inborn connections between cognitive and language brain modules. Tests and modeling should be used to understand how neural mechanisms of hearing enhance or suppress Helmholtz's dissonances originating in ear drums; is there scientific evidence for this Helmholtz's hypothesis? Why various species have different sensitivities to dissonances (or lack them at all). Which parts of musical ability are genetic and which are culturally developed?

The reviewed hypothesis suggests that language reduced direct connections between vocalization and ancient emotional centers. Neural imaging tests could reveal if music is connected to ancient emotional centers; is this connection direct? Is it different for music and language? To which extent and how does music involve emotional centers in cortex? Models of cultural evolution from section 8 should be extended to include the effects of music.

The reviewed hypothesis of the origins and functions of musical emotions addresses numerous questions, many of which remained opened for millennia. Therefore, small steps revealing neural mechanisms as well as studies of the suggested hypothesis about the function of music are necessary along with experimental laboratory tests, empirical ethnomusicological, anthropological, and historical studies. This review is a first step identifying the fundamental role of musical emotions in cognition and cultural evolution. Possibly it will form a foundation for a unified field of a multidisciplinary study. In conclusion, I would like to repeat that music is the most mysterious of human abilities, appealing directly to our primordial emotions, while connecting them to language and cognition.

4.13 Problems (* Indicates MS Level Problems; $^{+}$ Indicates PhD Level Problems)

P4.13.1 Interpret an Algorithm in Section 3.7 In mental Terms (Section 4.2)*

Select a problem similar to section 3.7 (using e.g. several hundred objects, known to the algorithms, and few dozens of situations, unknown to the algorithm). Write a computer simulation (say, using a MATLAB code) repeating an algorithm from section 3.7. Write an essay interpreting the algorithmic processes and results in mental terms (section 4.2): concepts, the knowledge instinct, emotions, imaginations, bottom-up and top-down signals, hierarchy (just two levels in this example), conscious and unconscious processes. Attempt to solve this problem using your favorite MATLAB clustering code. Describe differences in results. Describe difficulties if any.

P4.13.2 Use an Algorithm in Section 3.7 for Developing a System Learning a Simplified Form of Language with Bag-Models for Phrases (Section 4.3)*

Limit the system to two levels: words and phrases. Select from Internet (say, using Google) a data base consisting of large pieces of continuous text, about 1,000,000 words. Delete words 3 letter long or shorter. Find on the Internet rules of how to convert each word to its root (by pmitting "s" for plurals, "ed" for the past tense, etc. Simplify all 1,000,000 words to their roots. Select 30,000 words most often used in this data base. Model phrases using "bag models" (i) of no longer than 7 words; (ii) recognize that phrases are not necessarily consecutive words; non-essential words could intervene, as well as phrases could overlap; therefore allow a phrase to span 15 words. Using section 3.7 algorithm, learn 10,000 phrases that best describe the database. Write a plan of how to use these results to build as search engine for the Internet with elements of language understanding. (This problem can be used for a team of 2-3 writing several MS theses).

P4.13.3^{+} Use an Algorithm in Section 3.7 Along with the Dual Model in Section 4.4 for Developing a System Learning Objects and Their Names

Take a piece of a movie as a learning data. Select an educational movie for children, where a significant part of content is objects and their names. In every scene, or sequence of scenes recognize objects using an algorithm described in Problem P3.8.6. Use results of the previous problem P4.13.2 for recognition of words. Combine these results with the dual model 4.4.1; start again with vague models. (1) Start with vague models for objects and already learned crisp models for words. (2) Start with vague models for objects and for words. Compare results.

P4.13.4^{+} Continue the above Problem P4.13.3 for the Next Hierarchical Level of Abstract Concepts (Situations of Situations)

If words for some situations are not used in the movie, add required teaching episodes, like "this is a room." Alternatively, instead of a movie build a robot and teach it to understand words, objects, and situations in its environment.

P4.13.5$^+$ *Develop a multi-agent system learning language and cognition at many (adaptively varying) hierarchical levels in interaction among agents and with humans.*

P4.13.6$^+$ *Enhance the above multi-agent system by adding ability to learn relations between cognitive concepts at every level, markers for relating words, and language syntax.*

P4.13.7$^+$ *Enhance the above multi-agent system by adding more instinctual drives for values, corresponding emotions, behavioral actuators and behavioral models-representations necessary for a particular application (select yourself; certain drives, e.g. for energy, by charging batteries, for surviving/protecting oneself, will be necessary for any system.). Study evolution of cognitive dissonances; evolution of differentiation of KI (at every level) to resolve cognitive dissonances, synthesis of KI in the hierarchy, evolution of higher emotions, including the beautiful and sublime. Consider just one uninterrupted generation of agents.*

P4.13.8$^+$ *Use the above multi-agent system to study language evolution: growth in the number of words, in hierarchical levels, in evolution of grammar, syntax. Determine which parameters influence evolution of language emotionality. How emotionality affect evolution.*

Model genetic evolution of agents by using genetic algorithms. Model language transmission to next generations by each new agent learning language from surrounding agents and humans.

P4.13.9$^+$ *Enhance the above multi-agent system by adding musical ability: (1) add "inborn" connection between voice and emotions; (2) study evolution of music and emotions.*

Model evolution as above, by combining genetic and cultural evolution.

P4.13.10$^+$ *Study evolution of cultures; effects of music; what drives emotionality of languages? Types of languages and music?*

Model evolution as above, by combining genetic and cultural evolution.

4.14 Literature for Further Reading

4.14.1 Section 4.1, Fundamental Mind Mechanisms

Summaries and overviews of DL: Perlovsky 2001, 2006a, 2010, 2009c, 2010c,d.

4.14.2 Section 4.2, Dynamic Logic and Cognition

DL and cognition: Perlovsky, 1987, 1988, 2001, 2002, 2004, 2005, 2006a,b,c,d,e,f,g, 2007a,b, 2008, 2009a,b,c, 2010a,b,c,d,e,f,g; Perlovsky & McManus 1991; Fontanari & Perlovsky, 2007, 2008a,b; Fontanari et al, 2009; Ilin & Perlovsky, 2010; Perlovsky & Ilin, 2010a,b; Levine & Perlovsky, 2008; Perlovsky, Bonniot-Cabanac, & Cabanac, 2010.

Adaptive Resonance Theory (ART): Carpenter & Grossberg, 1987; Grossberg & Pearson, 2008; Grossberg & Versace, 2008.

Preliminary indications of the knowledge instinct in biology and psychology: Festinger, 1957; Berlyne, 1960, 1973; Harlow & Mears, 1979; Cacioppo & Petty, 1982.

Theory of instincts and emotions: Grossberg & Levine, 1987.

The knowledge instinct and aesthetic emotions: Perlovsky 1987, 1988; Perlovsky and McManus 1991; Perlovsky, 2001; 2002; 2006a; 2008, 2010a,b.

Experimental evidence for knowledge instinct: Perlovsky, Bonniot-Cabanac, & Cabanac 2009.

The knowledge instinct and higher cognitive functions: Perlovsky, 2002; 2006b,c,d,f; 2007a,b,d; 2008; 2009a,b,c; 2010a,b,d,e,f,h; Levine & Perlovsky, 2008; Perlovsky & Mayorga, 2007; Ilin & Perlovsky 2010; Perlovsky & Ilin, 2010a,b.

Purpose of life, meanings, beautiful and sublime emotions, and mathematical models of the mind: Levine & Perlovsky, 2008; Perlovsky, 2002, 2004, 2007b,d, 2009a,b,c.

Emotional intelligence: Cabanac, 2002; Mayer, 1999; Mayer, Salovey, & Caruso, 2008; Russell, 2003; Russell & Barrett, 1999; Spinoza, 2005; Tupes & Cristal, 1961.

Experimental data supporting mathematical theories: Bar et al, 2006; Perlovsky, Bonniot-Cabanac, & Cabanac, 2010; Festinger, 1957; Berlyne, 1960, 1973; Harlow & Mears, 1979; Cacioppo & Petty, 1982; Kosslyn, Ganis, & Thompson, 2001

Kantian aesthetics: Kant, 1790/1914, 1798/1974; Perlovsky, 2002, 2006c, 2008, 2010b.

History of philosophy and aesthetics: Aristotle, Topics; Kant, 1781/1943, 1790/1914, 1798/1974.

Contradictions in contemporary aesthetics: Perlovsky 2002, 2006a, 2006b, 2010a

Beautiful and NMF-DL: Perlovsky 2002, 2004, 2006a, 2006b, 2010a

The beauty of a scientific theory, quotes:
http://www.quotationspage.com/quote/26209.html; Poincare 1908;

4.14.3 Section 4.3, Natural Language Learning

Chomsky's linguistics: Chomsky 1965, 1972, 1981, 1995;

Cognitive linguistics: Croft & Cruse 2004; Evans & Green 2006; Feldman 2010; Ungerer & Schmid 2006.

Evolutionary linguistics: Hurford 2008; Christiansen & Kirby 2003; Brighton, Smith & Kirby 2005.

DL for language learning: Perlovsky 2004, 2006a,c, 2007a,b, 2009a; Fontanari & Perlovsky 2005, 2007b,c; Tikhanoff et al 2006; Perlovsky & Ilin 2010a,b.

4.14.4 Section 4.4, Integration of Language and Cognition

Language evolved on top of the system of mirror neurons: Arbib 2005.

Grounding: Meystel and Albus 2001.

Language Instinct: Pinker 1994.

DL, Dual model: Perlovsky 2006a,c, 2007c,d, 2009a,b; Perlovsky & Ilin 2010a,b.

Cognitive linguistics: Jackendoff 1983, 2002; Lakoff 1988; Lakoff & Johnson 1999; Langacker 1988; Talmy 1988, 2000; Kay 2002; Fauconnier & Turner 2008; Kay 2002.

Evolutionary linguistics: Christiansen & Kirby 2003; Christiansen & Chater 2008; Brighton et al 2005; Fontanari & Perlovsky, 2007; Fontanari et al 2009.

Inborn language mechanisms: Hauser, Chomsky, & Fitch 2002; Perlovsky 2007d.

Arbitrariness of vocalization: **Plato**

Models of vocal tract: Guenther 2006.

Supporting evidence: Arbib 2005; Franklin et al 2008; Deacon 1997; Mithen 1998; Bar et al, 2006; Levine & Perlovsky, 2008; Perlovsky, Bonniot-Cabanac, & Cabanac 2010.

Language abilities of primates: Savage-Rumbaugh & Lewine, 1994.

4.14.5 Section 4.5, Symbols: Grounded, Perceptual, and Amodal

What are symbols: Jung 1921; Deacon 1998; Peirce 1897, 1903; De Saussure 1916; Barsalou & Hale 1993; Perlovsky 2006b;d..

Perceptual symbols: Barsalou 1999, 2003a,b 2005, 2007, 2008; Simmons & Barsalou 2003; Yeh & Barsalou 2006; Kosslyn 1980; 1994; Perlovsky & Ilin 2011.

Mind and logic: Russell 1919; Hilbert 1928; Carnap 1959; Perlovsky 2001, 2006a, 2010g

Experimental evidence and future research: Wu & Barsalou 2009; Edelman & Newell 1998; Edelman 2003; Bar et al 2006; Bar 2007.

4.14.6 Section 4.6, Future Man-Machine Systems

Semantic web: Perlovsky 2006a,c, 2007d, 2009a,b; Perlovsky & Ilin 2010a.

4.14.7 Section 4.7, Emotional Intelligence and Love from the First Sight

Emotions: Cabanac 2002; Russell & Barrett 1999; Russell 2003; Juslin and Västfjäll 2008; Grossberg & Levine 1987; Spinoza 2005/1677; Dawkins 1976; Perlovsky 2006a,b, 2007a, 2009b, 2010g.
Intelligence components, "Big five": Tupes & Cristal 1961.
Emotional intelligence (EI): Mayer 1999; Mayer, Salovey, & Caruso 2008.
Emotional and conceptual intelligences: Jung 1921.

4.14.8 Section 4.8, Emotionality of Languages and Meanings

Animal vocalization: Deacon 1989; Lieberman 2000; Mithen 2007
Basic Emotions: Wikipedia http://en.wikipedia.org/wiki/List_of_emotions; Parrott 2001, Petrov et al 2011.
Language emotionality: Perlovsky 20097b,f, 2010g; Humboldt 1836/1967; Lerer 2007; Guttfreund 1990; Harris, Ayçiçegi, & Gleason 2003.

4.14.9 Section 4.9, Hierarchical Evolving Systems, the Beautiful and Sublime

Hierarchical system mathematical modeling: Perlovsky 1987; 1994; 1997; 1998; 2001; 2006a,b,c; 2007a,b,f; 2009b;2010g; Perlovsky, Plum, Franchi, Tichovolsky, Choi, & Weijers, 1997
Neuro-imaging data: Bar et al, 2006; Franklin et al, 2008
The beautiful: Perlovsky 2000,2002c,2010b,h.

4.14.10 Section 4.10, Evolution of Cultures

Evolution of cultures: Perlovsky 2007d, 2009b, 2010e,f; Perlovsky & Goldwag 2011; Humboldt 1836/1967.

4.14.11 Section 4.11, Emotional Sapir-Whorf Hypothesis

Sapir-Whorf hypothesis, SWH, (some people consider controversial the idea that thinking depends on language. Therefore we would emphasize that a person is not necessarily limited by his or her first language; by concentrating on thinking one can overcome limitations of language. However, as discussed in section 4.4., this is difficult, and most people are lead in their thinking by language): Bhartrihari IVCE/1971; Humboldt 1836/1967; Nietzsche 1876/1983; Sapir 1985; Whorf 1956; Roberson, Davidoff, & Braisbyb 1999;

Winawer, et al 2007; Wikipedia Sapir-Whorf Hypothesis, Linguistic relativity; Hurford 2008.

Emotional Sapir-Whorf Hypothesis: Perlovsky 2006d, 2009b, 2010a,g.

Experimental evidence: Guttfreund 1990; Balaskó & Cabanac 1998; Bar et al 2006; Barsalou 2009; Buchanan et al 2000; Harris, Ayçiçegi, & Gleason 2003; Franklin et al, 2008; Simmons et al, 2008.; Franklin et al 2008.

Akaike penalty function: Akaike 1974.

Preliminary indications of the knowledge instinct in biology and psychology: Festinger, 1957; Berlyne, 1960, 1973; Harlow & Mears, 1979; Cacioppo & Petty, 1982.

Language evolution: Botha 2003; Botha & Knight 2009; Hurford 2008; Fontanari et al 2005, 2007, 2008a,b, 2009; Mithen 2007.

Language faculty: Chomsky 1965, 1981, 1995; Hauser et al 2002.

Neural circuitry of vocalization: Deacon 1989; Lieberman 2000.

Neural circuitry of imagination: Kosslyn et al 2001.

Lagrangian formulation of physics: Feynman & Hibbs 1965.

Instinctual and emotional neural interactions: Gnadt and Grossberg 2008, Grossberg 1988; Grossberg & Levine 1987; Grossberg & Seidman 2006.

Music and emotions: Juslin & Västfjäll 2008.

Kant 1790/1914.

Evolution of English: Lerer 2007.

4.14.12 Section 4.12, Music: Its Function in Cognition and Evolution

Music is a mystery: Aristotle, IV BCE/1995, p.1434; Kant 1790; Darwin 1871; Juslin & Sloboda 2001; Masataka 2008; Pinker 1997; Editorial, 2008; Ball, 2008.

Western music theories from Pythagoras till 18th c.: James 1995; Plato 4c. BCE; Weiss & Taruskin 1984; Mattheson 1739; Avison 1753; Beattie 1778; Kant 1790.

Helmholtz theory of musical emotions: Helmholtz's 1863.

Current theories of musical emotions and empirical evidence: Frijda 1986; Meyer, Palmer, & Mazo 1998; Huron 1999; Juslin & Sloboda 2001; Sloboda & Juslin 2001; Justus & Hustler 2003; McDermott & Houser 2003; Trainor 2008; Fitch 2004; Livingstone & Thompson 2006; Cross 2008a,b; Cross & Morley 2008; Dissanayake 2000, 2008; Trehub 2003; Parncutt 2008; Levitin 2008; Juslin and Västfjäll 2008; Purwins et al 2008a,b; Coutinho & Cangelosi 2009.

Evolution of consciousness and cultures: Jaynes 1976; Diamond 1997; Perlovsky, 2006a,b,e,f, 2007, 2008; Levine & Perlovsky, 2008;.

Animal vocalizations: Darwin 1871; Deacon 1989; Davis et al 1996; Schulz et al 2005; Cross, 2008a; Masataka, 2008; Larson 1991; Seyfarth & Cheney 2003; Panksepp & Bernatzky, 2002; Trainor, 2008.

DL theory of musical emotions, differentiation and synthesis: Perlovsky 2005, 2006b,c,d,e,f, 2007, 2008a, 2009a,b,d, 2010a,i, 2011

Empirical evidence: Weiss and Taruskin 1984; Jaynes 1976; Simonton 1997; Perlovsky 2005, 2006b,e, 2008; Perlovsky, Bonniot-Cabanac, & Cabanac 2009.

Role of music in culture according to philosophers: Kant 1790; Nietzsche 1876; Jung 1921, 1934; Schopenhauer 1819.

Future experiments: Festinger, 1957; Akerlof & Dickens 2005; Juslin & Sloboda 2001; Steinbeis, Koelsch, & Sloboda 2006; Patel 2008;

Chapter 5
Epilogue Future Research Directions

Wide applicability of DL and performance gain achieved with it in solving practical problems, which could not have been solved previously, indicates that it is a fundamental mathematical result. Similarly, its wide applicability to cognitive phenomena, explanation and mathematical modeling of much that has not been understood previously, indicates that it is a fundamental mechanism of the mind-brain. Here we summarize the main ideas of the book: mathematical, cognitive, and future research directions.

5.1 Dynamic Logic: Mathematics, Engineering, and the Mind Summary

The main mathematical idea of DL is a process from vague to crisp. This process substitutes static statements of classical logic. For a logical statement to be applicable to real entities in the world, beyond an artificial world of axiomatically fixed meanings, the statement must be formulated as a DL statement-process. A statement here also means a model, plan, idea, concept (of understanding or behavior-action). Because our consciousness operates with nearly crisp mental states, similar to classical logic, our intuition is wired to classical logic, and formulating a problem in DL terms demands special effort, at least at the beginning. With little experience, DL can be used virtually for any problem, like calculus. DL requires formulating a problem as models that should match existing data. These models depend on parameters which values are not known, and which match the data with vagueness or fuzziness corresponding to uncertainty of parameters. At the end of the DL process, parameters approximate their true values, and models approximate patterns in the data.

Where the DL models are coming from? Should new models be developed for every new application? The DL model described in section 3.7 is a general method. It requires two steps, first, relationships among objects (or any entities) should be included into the model; and second, complex problems might require several hierarchical levels for modeling.

DL is a step beyond fuzzy logic. In fuzzy logic several fundamental operations, particularly fuzzification, de-fuzzification, learning, are performed using logical procedures, and in practical engineering applications they require human intervention. In DL these operations are combined in single dynamic-logic-process. The most important difference is that in DL the degree of fuzziness

L. Perlovsky et al.: Emotional Cognitive Neural Algorithms with Eng. Appl., SCI 371, pp. 175–198.
springerlink.com

(or vagueness) is automatically adapted during solution for multiple DL processes (models) running in parallel.

DL-NMF suggests approaches to mathematical modeling of a number of cognitive processes, which could not be modeled previously and some that had no explanation and seemed mysterious. These include perception, cognition, language, learning of situations, all previous attempts to model it mathematically have led to combinatorial complexity. DL is the only mathematical model that explains vagueness of mental representations at the initial stages of the perception process. It is a fundamental mechanism of DL. Moreover, DL predicted this more than a decade before it was discovered experimentally as a mechanism in the mind. DL suggests that it is a fundamental mechanism of the mind responsible for many mechanisms that could not have been formulated mathematically: bottom-up and top-down signal interactions at all hierarchical levels, hierarchical dynamics, interaction between language and cognition, music and its function in cognition, its evolution; evolution of cultures. As new mathematical modeling methods, these DL approaches have solved engineering problems unsolvable previously. As a fundamental mechanism of the mind it is a hypothesis that should be tested in psychological labs, and this testing by several research groups has began.

Recent DL predictions include (1) the initial stages ("gist") of higher level mental representations including context, situations, etc., is vague not only in terms of vagueness of constituting objects, but also in terms of vagueness of content (which objects belong or do not belong to a particular context or situation). (2) Dual hierarchical model of cognition and language explains many aspects of these abilities, which could not be previously understood, such as (2.1) how the brain-mind learns associations between words and objects; (2.2) extending this ability to higher levels (situations, abstract ideas, etc.); (2.3) why language is learned by 5 years of age, but cognitive understanding requires a lifetime. Predictions that can be easily verified in experiments are (2.4) existence of two connected mental representations for language and for cognition built on top of mirror-neuron system, (2.5) different vagueness of these representations, larger vagueness of higher level cognitive representations, and (2.6) this larger vagueness being less accessible to consciousness due to "masking" of cognitive representations by language ones.

5.2 Consciousness

DL gives a simple explanation and mathematical model of consciousness. Consciousness is an ability to concentrate attention on mental states. The more differentiated and concrete are the states the better they are accessible to consciousness. This is true about mental states representing understanding of the mind as well as about states representing understanding of the body. This eliminates "mystery" often associated with consciousness. Think how conscious you are about states of your stomach; as long as stomach works properly, we are not much conscious about it.

States of the mind are much more interesting and important for us than stomach. Yet, consciousness can be understood without mysteries. The principled difference is that bodily states are designed to function autonomously, unconsciously; but in the mind we strive for consciousness, for better understanding of the world and ourselves. Whereas we easily ignore unconscious bodily states, unconscious states of the mind fascinate us and seem more important than conscious one. The reason is that body usually is ruled by the law of negative feedback; this law minimizes distances between actual and desired states. This mechanism is not accessible to conscious. The mind works differently, it tries to match bottom-up and top-down signals; matched states are available to conscious. Let us repeat, the more differentiated and concrete are the states the better they are accessible to consciousness. This is true about mental states representing conceptual contents as well as about states representing emotions. People strongly differ in their modes of consciousness, some are more conscious about their conceptual states, other about their emotional states (Jung 1921). KI drives us toward more crisp and conscious states. Unconscious states of mind disturb us. We often pay more attention to what we do not understand than to what we understand. Therefore internal convictions of importance of mental states could be opposite to how conscious they are. We could value what we do not understand more than what we do. DL predicts that there is a corresponding difference in differentiating these states (details with which a person can describe contents of his or her mental states). This prediction can be easily tested in a psychological lab.

Interaction of language and cognition via the dual model predicts that higher level models (above directly perceptible objects) have to be much more conscious for language representations than for cognitive ones. This explains the famous discovery (Nobel Prize 2002) by Kahneman and Tversky of irrationality of human decision making. DL predicts that this irrationality is related to using language models instead of cognitive models. Language models are crisp and conscious, and therefore are easy to use. Contrary, cognitive models are vague and less conscious and therefore are difficult to use. Language models accumulate millennial cultural wisdom, but they are not based on personal real-life experience, and therefore they do not necessarily fit best the current personal situations or personal mode of consciousness. By relying on language models that are “good on average” people often make decisions that are opposite to their needs, irrational for their personal situations. This is the DL explanation for Tversky-Kahneman discovery. Exploration of this idea opens a whole new research direction in theory of human decision-making, rationality, language-cognition interaction, and conscious-unconscious. The above explanation-predictions, as well as future developments can be verified experimentally in psychological and neuro-imaging labs.

The above discussion has fundamental consequences for engineering decision-making systems. Whereas current systems rely on logical-linguistic models, future systems will model complex interactions of language-logical and cognitive DL models. These models might be less conscious in the minds of human decision makers; their modeling requires adaptive DL learning, and similarly their proper use by human analysts requires training aimed beyond verbal instructions.

There is no mystery about consciousness. Much confusion is related to two reasons. First is misunderstanding of operations of the dual model. To understand and model consciousness the dual hierarchical model is paramount. It explains why the same topic might appear at the same time as conscious (in language) and unconscious (in cognition). Second is related to operations of conscious and unconscious mechanisms of DL and PSS simulators. Whereas subjectively we feel conscious all the time unless we sleep, conscious states make up a tiny fraction of the mind mechanisms. We "feel" our mind as conscious and logical and tend to ignore majority of its unconscious states.

A great mystery has been historically associated with a particular aspect of consciousness: Free will. In the final count, do we have freedom, or are we automatons made of atoms and molecules, obeying physical laws?

5.3 Reductionism

"Reductionism" has been a fundamental difficulty in the past faced by scientists, philosophers, theologists, and anybody attempting possible scientific explanations of aesthetics or spiritual experiences, or phenomena of consciousness at higher levels of the mind, of free will. If a spiritual experience could be explained scientifically, it seemed the next step would be to reduce this explanation to biology, to chemistry, and to physics. The human being would be no different in principle than a rock, and the same fate would be faced by the beautiful and by God. Of course, most people would not tolerate this conclusion. But from a scientific logical viewpoint there was no escape from this conundrum. Some scientists therefore resorted to dualism (Descartes 1641, Spinoza 1677, Chalmers 1996), refusing to acknowledge that spirit and matter are of the same substance. Most scientists and theologists could not accept this solution since it contradicts the fundamental premises of monotheism and science. This conundrum seemed irresolvable.

The reductionism argument was a direct consequence of logic and logic was the foundation of science. There was though a huge hole in this line of reasoning: in the 1930s Gödel proved that logic is inconsistent, incomplete, and not as logical as expected. But scientists did not know how to use Gödel's results for resolving the problem of reductionism. Roger Penrose devoted two books to trying to connect the two and to escape reductionism of consciousness based on Gödel's arguments, but majority of scientists has not accepted his conclusions.

DL-NMF resolves this conundrum, not by parting with science or religion, but parting with the idea that logic is a fundamental mechanism of the mind. Instead of logic, we suggest, the fundamental mechanism of the mind is DL. To reiterate, DL is the process from vague to crisp. Most mind operations are vague, not logical; logical (or almost logical) thoughts, decisions, plans appear at the end of DL processes. This fact is hidden from our consciousness. Consciousness operates in such a way that we subjectively perceive our mind operations as purely conscious and logical. Our subjective intuition about the mind is based therefore on consciousness and logic. Yet, along with unconscious, the dynamic logic mechanism of the mind is confirmed in neuro-imaging experiments, and scientists and engineers will have to modify their intuitions.

The combination of vague and unconscious mechanisms eliminated the conundrum of reductionism. High level concepts involving the meaning of life, beautiful and sublime are vague and unconscious. We can analyze them and study involved neural mechanisms scientifically. But these high level concepts cannot be reduced to finite combinations of constituent simpler concepts. This argument is related to dynamic logic, to solving the problem of computational complexity, and to the Gödel theory. Computational complexity is due to the fact that high-level decisions involve a choice from a near infinite number of combinations of lower-level concepts. Therefore these decisions involve near infinite information.

The seemingly unsolvable conundrum of reductionism, which has led many people to doubts about the possibility of combining science, consciousness, aesthetics, and religion, others to dualism, or to postulating future non-computable science, is resolved now. It has become clear that these doubts were based on wrong intuition, on assuming that the mind's main mechanism is logic, that the mind moves in time smoothly from one conscious logical state to another. We know now that conscious logical states of mind are tiny islands among non-logical and unconscious operations, processes of dynamic logic. Freedom of will creates a contradiction in logic, but there is no contradiction about free will in the mind.

5.4 Making a Scientific Revolution

Why are some important mathematical discoveries immediately recognized and adopted by engineering community, such as e.g. Aristotelian logic, and logic-based AI, whereas other immensely important discoveries remain misunderstood and unaccepted for years; these include Aristotelian theory of mind, the Gödelian theory (recognized overnight, but implications are still ignored), Zadeh's fuzzy logic, and others? One may wonder why, despite the Gödel's theory developed in the 1930s and immediately recognized as a fundamental result, mathematicians still relied on formal logic when developing artificial intelligence in the 1950s and 60s, and many still rely today?

We would emphasize that this topic is essential for improving success of the entire scientific and engineering enterprise. Engineering and scientific community used to relegate these questions to "philosophy," unneeded to engineers, or belonging at best to marketing. This section suggests that existing knowledge of the mind and its models are ready to consider this question as an essential part of science and engineering. The novel research direction proposed here considers acceptance (or not) of scientific ideas as a scientific topic studying processes in the mind-brain, and therefore being a subject for study.

Processes of accumulation of scientific knowledge, changes of scientific paradigms, "scientific revolutions" have not been studied by scientists, but were studied by philosophers, especially in the 20th c. A traditional view on growth of knowledge was that empirical observations accumulate and are subsequently generalized. Karl Popper repudiated this classical view that science grows by inductions from observations. He suggested that scientific knowledge grows by advancing multiple scientific hypotheses, among which most are later empirically falsified; those that survived become scientific theories, until they are falsified and new theories are advanced in this process.

This however is not true. Science does not grow by generating random hypotheses and then falsifying them. Newton specifically wrote that he does not advance hypotheses. Instead, as we can understand his and many other scientists' thinking process, scientific thinking is directed by intuitions. These intuitions have to correspond to a large amount of knowledge and experimental data existing in every field. Coming up with even a single hypothesis explaining the wealth of experimental data is a rare event. Scientific knowledge therefore grows not by a routine procedure of falsification of wrong hypotheses, but by creative process of scientific intuition, which creates new scientific ideas. A new scientific idea should, in addition to explaining existing data, make experimentally testable predictions. Until these predictions are tested and confirmed, it is customary to call the idea a hypothesis; it is acknowledged as a valid theory as its predictions are gradually confirmed. Einstein, Poincare, and some other great scientists, however, considered a first proof of validity of a scientific theory its beauty. According to DL, the beauty of a theory is similar to other aesthetic emotions at the top of the mind hierarchy. It is related to the emotional feel of purpose. A theory is purposeful and meaningful if it explains knowledge in a wide field with few assumptions.

Thomas Kuhn analyzed the process of scientific revolutions, the process in which a previously acknowledged theory is substituted by a new one. He emphasized that this process is not as clear-cut as experimentally proving predictions of a new theory. He analyzed historical processes of how new theories are accepted. He found that recognized experts are not going to acknowledge that they were wrong, just because some measurements, which they cannot explain, support a new theory. There are always reasons for doubts about a new theory and supporting data. A new theory, Kuhn wrote, even if valid and beautiful, will only be accepted after recognized experts will retire, and a new generation of scientists, those that grew up along with the new theory, will come to occupy University chairs. The actual process could take longer than a generation, since the new generation are students of retired experts, receiving knowledge from old hands, and may tend to continue rejecting new ideas.

Nobody so far investigated which properties of new theories make them readily acceptable, whereas other, no less fundamental ideas wait long time to be accepted. Because of importance of this subject for the entire science and engineering discipline, this should become a future field of study.

DL suggests that properties of consciousness, its logical bias discussed in previous sections, influence acceptance of new theories. This conscious-logical "bias" affects, which theories are accepted and which are ignored for long time. Theories relying on conscious, logical mechanisms are accepted faster. Returning to the beginning of this section, logical bias of consciousness explains why, despite the Gödel's theory, mathematicians still relied on formal logic when developing artificial intelligence in the 1950s and 60s, and many still rely on logical rules today. Logic-based AI was immediately accepted. On the other hand, theories relying on unconscious and illogical mechanisms are accepted only after many years. Among those are Zadeh's fuzzy logic, Kanehman-Tversky's theory (2002 Nobel Prize after Tversky died), Grossberg's theories of neural mechanisms of the mind. The conclusion from the above analysis is that theories of illogical mechanisms remain misunderstood and unaccepted for years because of logical bias in scientific thinking.

5.5 Science and Religion

5.5.1 Why Adam Was Expelled from Paradise, Cognitive Science View

5.5.1.1 KI and Heuristics

Using KI for making decisions is not the only way of thinking. Effort minimization (EM) is an alternative long-established biological principle. When applied to thinking, it suggests that people make decisions by relying on heuristics learned from parents, friends, and from surrounding culture, rather than by using KI. Heuristics contain millennial wisdom, they are formulated as ready-made rules in language, and can be used fast, without much thinking. But they may not fit to concrete individual situations. Developers of artificial intelligence in the 1960s and 70s attempted to model human decision making using heuristics, but this effort failed: adaptation to concrete conditions is essential.

From the work of the pioneering 18th century mathematicians Jakob Bernoulli and Thomas Bayes through the late 20th century, the dominant notion in the psychology of human decision making was based on rational optimization. The belief was that each decision maker had an internal, and self-consistent, *subjective utility function*, and made all choices involving risk by choosing the alternative for which the mathematical expectation of utility was the largest. But all that changed with the work, starting in the late 1960s, of Daniel Kahneman, winner of the 2002 Nobel Prize in economics, and Amos Tversky, who would have shared that prize had he been alive (Tversky and Kahneman 1974, 1981).

Tversky and Kahneman found that in many choices relating to gain and loss estimation, preferences run counter to rational optimization and lack self-consistency over different linguistic framings of the choice. For example, subjects asked to consider two programs to combat an Asian disease expected to kill 600 people tend to prefer the certain saving of 200 people to a 1/3 probability of saving all 600 with 2/3 probability of saving none. However, subjects also tend to prefer a 1/3 probability of nobody dying with a 2/3 probability of 600 dying to the certainty of 400 dying. The choices are identical in actual effect, but are perceived differently because of differences in frame of reference (comparing hypothetical states in one case with the state of all being alive, in the other case with the state of all dying). Tversky and Kahneman explain their data by noting that "choices involving gains are often risk averse while choices involving losses are often risk taking" (Tversky and Kahneman, 1981, 453).

Heuristics have evolutionary value despite sometimes leading to errors and information losses. Heuristic simplification is particularly useful when a decision must be made rapidly on incomplete information, or when the stakes of the decision are not high enough to justify the effort of thorough deliberation. An example is buying a box of cereal in a supermarket (Levine 1997).

5.5.1.2 Adam and Eve

The origin of the controversy between the KI and heuristics can be traced to the first pages of the Bible, to the story of Adam and Eve. In the 12th century Moses Maimonides, in his "Guide for the Perplexed" (Maimonides 1190/1956) analyzed the relationship between KI and heuristics. He was asked by his student: "Why did God, on one hand, give Adam the mind and free will, while on the other, forbid him to eat of the tree of knowledge? Did God not want Adam to use his mind?" Maimonides answered that God gave Adam the mind to think for himself what is good and what is bad (we associate this ability with the KI). But Adam succumbed to temptation and ate from the tree of knowledge. Adam thereby took a "shortcut" and acquired ready-made heuristics, that is, rule-of-thumb knowledge to guide him so his choices did not require hard thinking. In conclusion, Maimonides explained that Adam's story described our predicament. Whereas God's ultimate commandment is to use the KI, it is difficult and we are not completely capable of doing it, especially when thinking about the highest values. Adam's story described the workings of our mind: struggle between the KI and EM. EM provides the surety of millennial cultural support, but may not suit your individual circumstances. The KI may lead to doubts and uncertainties, but if successfully used, leads to the satisfaction of being more conscious about your choices.

Maimonides' interpretation of the Biblical story adds another dimension to the previously discussed differences between the KI and EM. Mathematically, it is possible to formulate a minds' utility function so that the KI and EM are brought close to each other. This utility function can account for the survival value of quick decisions and also for the limited amount of any individual experience, for uncertainty in observation of data, and for minimizing the worst-case losses (such as preventing death) versus maximizing average gain. The utility function even can account for the fact that future is unknown and therefore individual experience should be integrated with culturally accumulated knowledge. But Maimonides hints at something different, something more fundamental than correct formulation of a utility function. He suggests that "original sin" determining the basic imperfectness of humankind is related to how we do or do not use our ability for knowledge and for making conscious choices.

In summary, the choice between increase of knowledge and minimization of cognitive effort, between the KI and EM, Maimonides connected to original sin. The Bible identifies it as the "fallen" condition of the mankind, the source of the world's miseries. Buddhism sees the source of human unhappiness as *tanha*, loosely translated as "desire" or "attachment" (Smith 1958) but, from a scientific perspective, meaning self-absorbed deficiency-based emotions leading to over-reliance on EM heuristics. The KI involves individual effort for increasing knowledge, aesthetic emotions; at the highest levels of the mind hierarchy it involves the beautiful and sublime. It also involves the conscious and the unconscious, the conceptual and the emotional, language and thinking. There is a difference between the "fallen," bodily emotions involved in EM, in using language without thinking, and aesthetic emotions related to the KI.

The higher up we go in the hierarchy of mind, the closer we are to the beautiful and sublime, the easier, it seems, to succumb to the temptation to stop thinking, to stop using cortex, the uniquely human brain region, and to use ready-made concepts acquired from the culture: language concepts connected not to individual thinking, concepts connected to "Mom and Dad prohibitions," to amygdalar emotions triggered by previous failures, when we tried to think and got burned.

To summarize, Maimonides suggested that Adam was expelled from paradise for refusing to think for himself. This Maimonides' explanation we connected to how humans use KI and EM (Levine & Perlovsky 2008).

5.5.2 Religion from Scientific Point of View

"Everyone who is seriously involved in the pursuit of science becomes convinced that a spirit is manifest in the laws of the Universe." This Einsteinian statement remains outside of science. Connecting science with the highest spiritual quests of the human mind is essential for continuation of culture. Carl Jung wrote that schism between science and religion points to a psychosis of contemporary collective psyche; survival of culture demands repairing this schism. Many outstanding scientists and theologists attempt this. Many books are written arguing that scientific discoveries do not contradict the main tenets of the world's religions. Yet, there has been no unifying approach, science and religion remained in two separate parts of the mind. There has been no bridge between the two; no scientific approach to spiritual dimensions of the mind-brain. With the knowledge instinct and DL, science approaches mechanisms of human spiritual abilities.

Teleology explains the Universe in terms of purposes. In many religious teachings, it is a basic argument for the existence of God: If there is purpose, an ultimate Designer must exist. Therefore, teleology is a hot point of debates between creationists and evolutionists: Is there a purpose in the world? "Evolutionists" believing in evolution assume that the only explanation is causal. Newton laws gave a perfect causal explanation for the motion of planets: A planet moves from moment to moment under the influence of a gravitational force. Similarly, today science explains motions of all particles and fields according to causal laws, and there are exact mathematical expressions for fields, forces and their motions. Causality explains what happens in the next moment as a result of forces acting in the previous moment. Most scientists accept this causal explanation and oppose to teleological explanations in terms of purposes. The very basis of science, it seems, is on the side of causality, and religion is on the side of teleology.

However, at the level of the first physical principles this is not so. The contradiction between causality and teleology does not exist at the very basic level of fundamental physics. The laws of physics, from classical Newtonian laws to quantum superstrings, can be formulated equally as causal or as teleological. An example of teleological principle in physics is energy minimization, particles move so that energy is minimized. As if particles in each moment know their purpose: to minimize the energy. The most general physical laws are formulated as minimization of Lagrangian. Lagrangian is a more general physical entity than energy. Causal dynamics, motions of particles, quantum strings, and superstrings are determined by minimizing Lagrangian (Feynman & Hibbs 1965). A particle under force moves from point to point as if it knows its final purpose, to minimize Lagrangian. Causal dynamics and teleology are two sides of the same coin.

DL and the knowledge instinct are mathematically similar to Hamiltonian and Lagrangian formulations of general physical laws: evolution of the mind is guided by causal dynamics (DL), which is equivalent to a teleological principle of knowledge maximization. In this regards the KI is a revolutionary principle. For the first time it states that for a very complex system, the human mind, causality and purpose are equivalent. Instead of rule of entropy and thermal death, the human destiny is ruled by increase of knowledge. The knowledge instinct defines a new "arrow of time". One does not have to choose between scientific explanation and teleological purpose: Causal scientific dynamics and purpose-driven dynamics (teleology) are mathematically equivalent.

Scientific understanding of the beautiful and sublime corresponds to artistic and teleological ones: these are not final notions that could be formulated axiomatically. We discussed in details in section 5.3 that science is not reducible. Mechanisms of the highest aspirations of human spirit are not logically reducible to finite statements. Attempts to compute them logically exceed in complexity all elementary interactions in the Universe in its entire lifetime and therefore logical choices of beautiful and sublime involve more information than is available in the Universe. A possibility of these choices is called a miracle in traditional language. DL gives a computational theory of these choices without reducibility.

Analyzing beautiful in section 4.1 we concluded that it is an emotion of satisfaction of KI at the highest levels of the mind hierarchy; every step toward understanding the meaning and purpose of our existence we feel as beautiful. But conceptual understanding is not sufficient, the knowledge instinct also strives for understanding behavior, for actions that would realize the beautiful in our life. Every step toward this is experienced emotionally as spiritually sublime feelings. DL suggests that mental representations at the top of our mind unify our entire experience and are perceived as the meaning and purpose of our existence. These representations are vague and unconscious. They do not belong to our conscious I. They are outside of our consciousness.

In our culture, since the ascendance of science, many people consider themselves non-religious. But it is not in one's power to change the unconscious structure of the mind. The representation of our highest purposiveness is outside of our conscious control. The scientific analysis in this book leads to a conclusion that it is not in our power to be "religious" or "irreligious." One could participate in an organized religion or refuse to do so. One could consider himself or herself a non-religious person. Or one could choose to study what is known about the contents of the highest models from accumulated wisdom of theologists and philosophers, or by combining this wisdom with the scientific method, as the science-and-religion community does. One can choose to refer to the agency property of the unconscious model at the top of the mind hierarchy, and yet refuse or accept to use the word God.

Understanding of the mind mechanisms today came close to bridging spirituality and science. Religious principles can be understood scientifically, by understanding human mind. Contents of models of beautiful and sublime are unconscious; they do not belong to our consciousness. They are "collective," outside of consciousness. Consciousness does not control them, *they* control individual consciousness. Therefore, we feel them as a source of agency outside of ourselves. In traditional cultures and among religious people throughout the world this source of agency is called God, in recent arguments it is called Designer.

5.6 Problems (*MS Level Problems; +PhD Level Problems)

5.6.1+ Experimentally verify the DL prediction: the initial stages ("gist") of higher level mental representations including context, situations, etc., are vague not only in terms of vagueness of constituting objects, but also in terms of vagueness of content (which objects belong or do not belong to a particular context or situation).

5.6.2+ Experimentally verify the dual model prediction: existence of two types of connected mental representations for language and for cognition built on top of mirror-neuron system.

5.6.3+ Experimentally verify the DL and dual model prediction: language representations are crisper and more conscious than cognitive representations (especially at higher levels).

5.6.4+ Experimentally verify the DL and dual model prediction: when talking (or silent reading) the ratio of excitation of cognitive brain areas relative to language brain areas goes down, when talking/reading about abstract, high-level ideas.

5.6.5+ Experimentally verify the following: better differentiated cognitive states (understood in more details) are more conscious and less emotional than less understood and less differentiated.

5.7 Literature for Further Reading

DL reviews: Perlovsky, 2001; 2006a; 2010c,g.
Consciousness: Perlovsky, 2010g,j; Bar 2006, Grossberg, 1999; Descartes, 1641; Spinoza, 1677; Chalmers, 1996; Jung, 1921.
Heuristics: Tversky & Kahneman 1974; 1981; Levine & Perlovsky, 2008.

Evolution of science: Popper, 2002; Kuhn 1962; Lakatos & Musgrave,1965.
Science and religion: Maimonides 1190, Levine & Perlovsky 2008, Perlovsky 2010e,j, Feynman & Hibbs, 1965.

Acknowledgments

We are thankful to M. Alexander, M. Bar, R. Brockett, M. Cabanac, A. Cangelosi, F. Fontanari, F. Lin, J. Gleason, B. Kovalerchuk, R. Kozma, D. Levine, A. Ovsich, I. Ternovskyi, and B. Weijers, and to AFOSR PMs J. Sjogren, D. Cochran, J. Zhang, and A. Nachman for supporting part of this research.

Related Web Pages

http://www.leonid-perlovsky.com/
L.I. Perlovsky, SciTopic Web Encyclopedia articles
Beauty and art. Cognitive function and evolution;
http://www.scitopics.com/Beauty_and_art_Cognitive_function_and_evolution.html
Consciousness: a possibility of free will. Cognitive science and mathematical model of the mind.
http://www.scitopics.com/Consciousness_a_possibility_of_free_will.html
Dynamic logic; http://www.scitopics.com/Dynamic_logic.html
Dynamic logic, computational complexity, engineering and mathematical breakthroughs;
http://www.scitopics.com/Dynamic_logic_computational_complexity_engineering_and_mathematical_breakthroughs.html
Dynamic Logic. Cognitive Breakthroughs;
http://www.scitopics.com/Dynamic_Logic_Cognitive_Breakthroughs.html
EM, Dynamic Logic, Gaussian mixtures, cognitive algorithms, and brain mechanisms;
http://www.scitopics.com/EM_Dynamic_Logic_Gaussian_mixtures_cognitive_algorithms_and_brain_mechanisms.html
Jihad and Arabic Language. How Cognitive Science Can Help Us Understand the War on Terror;
http://www.scitopics.com/Jihad_and_Arabic_Language_How_Cognitive_Science_Can_Help_Us_Understand_the_War_on_Terror.html
Language and Cognition. Interaction Mechanisms;
http://www.scitopics.com/Language_and_Cognition_Interaction_Mechanisms.html
Languages and Cultures: Emotional Sapir-Whorf Hypothesis (ESWH); http://www.scitopics.com/Languages_and_Cultures_Emotional_Sapir_Whorf_Hypothesis_ESWH.html
Mind mechanisms: concepts, emotions, instincts, imagination, intuition, beautiful, spiritually sublime;
http://www.scitopics.com/Mind_mechanisms_concepts_emotions_instincts_imagination_intuition_beautiful_spiritually_sublime.html

Music and Emotions. Functions, Origins, Evolution;
http://www.scitopics.com/Music_and_Emotions_Functions_Origins_Evolution.html
New "Arrow of Time": Evolution increases knowledge;
http://www.scitopics.com/New_Arrow_of_Time_Evolution_increases_knowledge.html
Physics of the mind: Concepts, emotions, language, cognition, consciousness, beauty, music, and symbolic culture;
http://www.scitopics.com/Physics_of_the_mind_Concepts_emotions_language_cognition_consciousness_beauty_music_and_symbolic_culture.html
Science and Religion: New cognitive findings bridge the fundamental gap;
http://www.scitopics.com/Science_and_Religion_New_cognitive_findings_bridge_the_fundamental_gap.html.

References and Bibliography

Akaike, H.: A new look at the statistical model identification. IEEE Transactions on Automatic Control 19(6), 716–723 (1974)

Akerlof, G.A., Dickens, W.T.: The economic consequences of cognitive dissonance. In: Akerlof, G.A. (ed.) Explorations in Pragmatic Economics. Oxford University Press, New York (2005)

Arbib, M.A.: From monkey-like action recognition to human language: An evolutionary framework for neurolinguistics. Behavioral and Brain Sciences 28, 105–167 (2005)

Aristotle: The complete works. The revised Oxford translation: Barnes, J. (ed.) Princeton Univ. Press, Princeton (Original work IV BCE) (1995)

Balaskó, M., Cabanac, M.: Grammatical choice and affective experience in a second-language test. Neuropsychobiology 37, 205–210 (1998)

Ball, P.: Facing the music. Nature 453, 160–162 (2008)

Bar, M.: Predictions: A universal principle in the operation of the human brain (Introduction). Theme issue: Predictions in the brain: Using our past to generate a future. In: Bar, M. (ed.) Philosophical Transactions R. Soc., vol. B364, pp. 1181–1182 (2009)

Bar, M.: The proactive brain: using analogies and associations to generate predictions. Trends in Cognitive Sciences 11(7), 281–289 (2007)

Bar, M., Kassam, K.S., Ghuman, A.S., Boshyan, J., Schmid, A.M., Dale, A.M., Hämäläinen, M.S., Marinkovic, K., Schacter, D.L., Rosen, B.R., Halgren, E.: Top-down facilitation of visual recognition. Proceedings of the National Academy of Sciences USA 103, 449–454 (2006)

Barrett, L.F., Bar, M.: See it with feeling: Affective predictions in the human brain. Royal Society Phil. Trans. B 364, 1325–1334 (2009)

Barsalou, L.W.: Simulation, situated conceptualization, and prediction. Phil. Trans. R. Soc. B 364, 1281–1289 (2009)

Barsalou, L.W., Hale, C.R.: Components of conceptual representation: From feature lists to recursive frames. In: Van Mechelen, I., Hampton, J., Michalski, R., Theuns, P. (eds.) Categories and Concepts: Theoretical Views and Inductive Data Analysis. Academic Press, London (1993)

Berlyne, D.E.: Conflict, Arousal, and Curiosity. McGraw-Hill, New York (1960)

Bhartrihari: The Vâkyapadîya, Critical texts of Cantos I and II with English Translation. Pillai, K.(trans.) Motilal Banarsidass, Delhi (1971) (IVCE)

Botha, R.F.: Unraveling the Evolution of Language. Elsevier, Amsterdam (2003)

Botha, R.F., Knight, C. (eds.): The Cradle of Language. Oxford University Press, New York (2009)

Brighton, H., Smith, K., Kirby, S.: Language as an evolutionary system. Phys. Life Rev. 2(3), 177–226 (2005)

Buchanan, T.W., Lutz, K., Mirzazade, S., Specht, K., Shah, N.J., Zilles, K., et al.: Recognition of emotional prosody and verbal components of spoken language: an fMRI study. Cognitive Brain Research 9, 227–238 (2000)

Cabanac, M.: What is emotion? Behavioural Processes 60, 69–84 (2002)

Cacioppo, J.T., Petty, R.E.: The need for cognition. Journal of Personality and Social Psychology 42, 116–131 (1982)

Cacioppo, J.T., Petty, R.E., Feinstein, J.A., Jarvis, W.B.G.: Dispositional Differences in Cognitive Motivation: The Life and Times of Individuals Varying in Need for Cognition. Psychological Bulletin 119(2), 197–253 (1996)

Cangelosi, A., Greco, A., Harnad, S.: From robotic toil to symbolic theft: grounding transfer from entry-level to higher-level categories. Connect. Sci. 12, 143–162 (2000)

Cangelosi, A., Riga, T.: An embodied model for sensorimotor grounding and grounding transfer: experiments with epigenetic robots. Cogn. Sci. 30, 673–689 (2006)

Cangelosi, A., Bugmann, G., Borisyuk, R. (eds.): Modeling Language, Cognition and Action: Proceedings of the 9th Neural Computation and Psychology Workshop. World Scientific, Singapore (2005)

Cangelosi, A., Parisi, D. (eds.): Simulating the Evolution of Language. Springer, London (2002)

Cangelosi, A., Tikhanoff, V., Fontanari, J.F., Hourdakis, E.: Integrating language and cognition: A cognitive robotics approach. IEEE Computational Intelligence Magazine 2(3), 65–70 (2007)

Carnap, R.: The Logical Syntax of Language. Littlefield, Adams & Co., Totowa (1959)

Carpenter, G.A., Grossberg, S.: A massively parallel architecture for a self-organizing neural pattern recognition machine. Computer Vision, Graphics, and Image Processing 37, 54–115 (1987)

Chalmers, D.: The conscious mind: In search of a fundamental theory. Oxford University Press, NewYork (1997)

Chomsky, N.: Aspects of the theory of syntax. MIT Press, Cambridge (1965)

Chomsky, N.: Principles and parameters in syntactic theory. In: Hornstein, N., Lightfoot, D. (eds.) Explanation in Linguistics the Logical Problem of Language Acquisition. Longman, London (1981)

Chomsky, N.: The minimalist program. MIT Press, Cambridge (1995)

Christiansen, M.H., Chater, N.: Language as shaped by the brain. Behavioral and Brain Sciences 31(5), 489–509 (2008)

Christiansen, M.H., Kirby, S.: Language evolution. Oxford Univ. Press, New York (2003)

Confucius. Analects. In: Lau, D.C. (tr.) The Chinese University Press, Hong Kong (2000) (551–479 B.C.E.)

Coutinho, E., Cangelosi, A.: The use of spatio-temporal connectionist models in psychological studies of musical emotions. Music Perception 27(1), 1–15 (2009)

Coventry, K.R., Lynott, L., Cangelosi, A., Knight, L., Joyce, D., Richardson, D.C.: Spatial language, visual attention, and perceptual simulation. Brain and Language (2009) (in press)

Croft, W., Alan Cruse, D.: Cognitive linguistics. Cambridge University Press, Cambridge (2004)

Cross, I.: The evolutionary nature of musical meaning. Musicae Scientiae, 179-200 (2008a), http://www.mus.cam.ac.uk/%7Eic108/PDF/IRMC_MS07_1.pdf

Cross, I.: Musicality and the human capacity for culture. Musicae Scientiae, Special issue, 147–167 (2008b)

Cross, I., Morley, I.: The evolution of music: theories, definitions and the nature of the evidence. In: Malloch, S., Trevarthen, C. (eds.) Communicative Musicality, pp. 61–82. Oxford University Press, Oxford (2008)

Darwin, C.R.: The descent of man, and selection in relation to sex. John Murray, London (1871)

Davis, P.J., Zhang, S.P., Winkworth, A., Bandler, R.: Neural control of vocalization: respiratory and emotional influences. J. Voice 10, 23–38 (1996)

Deacon, T.W.: The neural circuitry underlying primate calls and human language. Human Evolution 4(5), 367–401 (1989)

Deacon, T.W.: The symbolic species: the co-evolution of language and the brain. Norton, New York (1997)

Deming, R., Perlovsky, L.I.: A Mathematical Theory for Learning, and its Application to Time-varying Computed Tomography. New Math. and Natural Computation 1(1), 147–171 (2005)

Deming, R.W., Perlovsky, L.I.: Concurrent multi-target localization, data association, and navigation for a swarm of flying sensors. Information Fusion 8, 316–330 (2007)

Deming, R., Schindler, J., Perlovsky, L.: Multitarget/Multisensor Tracking using only Range and Doppler Measurements. IEEE Transactions on Aerospace and Electronic Systems 45(2), 593–611 (2009)

Descartes, R.: Meditations on First Philosophy, in The Philosophical Writings of René Descartes (1641); Cottingham, J., Stoothoff, R., Murdoch, D.(trans.) vol. 2, pp. 1–62. Cambridge University Press, Cambri dge (1984)

Diamond, J.: Guns, germs, and steel: The fates of human societies. W.W. Norton, & Co., New York (1997)

Dissanayake, E.: Antecedents of the temporal arts in early mother-infant interactions. In: Wallin, N., Merker, B., Brown, S. (eds.) The origins of music, pp. 389–407. MIT Press, Cambridge (2000)

Dissanayake, E.: If music is the food of love, what about survival and reproductive success? Musicae Scientiae Special Issue, 169-195 (2008)

Duda, R., Hart, P., Stork, D.: Pattern Classification, 2nd edn. Editorial. Wiley Interscience, Hoboken (2000); Bountiful noise. Nature 453, 134 (2008)

Endsley, M.R.: Toward a theory of situation awareness in dynamic systems. Human Factors 37(1), 32–64 (1995)

Evans, V., Green, M.: Cognitive linguistics: an introduction. Edinburgh University Press, Edinburgh (2006)

Fauconnier, G., Turner, M.: The origin of language as a product of the evolution of modern cognition. In: Laks, B., et al. (eds.) Origin and Evolution of Languages: Approaches, Models, Paradigms. Equinox, London (2008)

Feldman, J.: Embodied language, best-fit analysis, and formal compositionality. Physics of Life Reviews 7(4), 385–410 (2010)

Festinger, L.: A Theory of Cognitive Dissonance. Stanford University Press, Stanford (1957)

Feynman, R., Hibbs, R.A.: Quantum Mechanics and Path Integrals. McGraw-Hill, New York (1965)

Fiske, J.: Film, TV and the Popular. In: Bell, Hanet (eds.) Continum: The Australian Journal of Media & Culture, vol. 12, pp. 4–5 (1987)

Fitch, W.T.: On the biology and evolution of music. Music Perception 24, 85–88 (2004)

Fontanari, J.F., Perlovsky, L.I.: Meaning Creation and Modeling Field Theory. In: IEEE Int. Conf. Integration of Knowledge Intensive Multi-Agent Systems, Waltham, MA (2005)

Fontanari, J.F., Perlovsky, L.I.: Evolving Compositionality in Evolutionary Language Games. IEEE Transactions on Evolutionary Computations 11(6), 758–769 (2007), doi:10.1109/TEVC.2007.892763

Fontanari, J.F., Perlovsky, L.I.: A game theoretical approach to the evolution of structured communication codes. Theory in Biosciences 127(3), 205–214 (2008a)

Fontanari, J.F., Perlovsky, L.I.: How language can help discrimination in the Neural Modeling Fields framework. Neural Networks 21(2-3), 250–256 (2008b)

Fontanari, J.F., Tikhanoff, V., Cangelosi, A., Perlovsky, L.I., Ilin, R.: Cross-situational learning of object-word mapping using Neural Modeling Fields. Neural Networks (2009)

Franklin, A., Drivonikou, G.V., Bevis, L., Davie, I.R.L., Kay, P., Regier, T.: Categorical perception of color is lateralized to the right hemisphere in infants, but to the left hemisphere in adults. PNAS 105(9), 3221–3225 (2008)

Frijda, N.H.: The emotions. Cambridge University Press, Cambridge (1986)

Garvin, L., Perlovsky, L.I.: Statistical Quantum Theory and Statistical Pattern Recognition. In: Pandalai (ed.) Current Topics in Pattern Recognition Research, Research Trends, Trivandrum, India (1995)

Gibson, J.J.: The ecological approach to visual perception. Houghton Mifflin, Boston (1979)

Glenberg, A.M.: What memory is for. Behavioral & Brain Sciences 20, 1–55 (1997)

Glenberg, A.M., Kaschak, M.: Grounding language in action. Psychonomic Bulletin & Review 9(3), 558–565 (2002)

Gnadt, W., Grossberg, S.: SOVEREIGN: An autonomous neural system for incrementally learning planned action sequences to navigate towards a rewarded goal. Neural Networks 21, 699–758 (2008)

Gödel, K.: Collected works. In: Feferman, S., Dawson Jr., J.W., Kleene, S.C. (eds.) Publications 1929-1936, vol. I. Oxford Univ. Press, New York (1931/1994)

Grossberg, S.: Neural networks and natural intelligence. MIT Press, Cambridge (1988)

Grossberg, S.: The Link between Brain Learning, Attention, and Consciousness. Consciousness and Cognition 8, 1–44 (1999)

Grossberg, S., Levine, D.S.: Neural dynamics of attentionally modulated Pavlovian conditioning: blocking, inter-stimulus interval, and secondary reinforcement. Psychobiology 15(3), 195–240 (1987)

Grossberg, S., Seidman, D.: Neural dynamics of autistic behaviors: Cognitive, emotional, and timing substrates. Psychological Review 113, 483–525 (2006)

Grossberg, Pearson (2008)

Grossberg, Versace (2008)

Guenther, F.H.: Cortical interactions underlying the production of speech sounds. Journal of Communication Disorders 39, 350–365 (2006)

Guttfreund, D.G.: Effects of language usage on the emotional experience of Spanish-English and English-Spanish bilinguals. J. Consult. Clin. Psychol. 58, 604–607 (1990)

Hall, D.L., Llinas, J.: Handbook of multisensory data fusion. CRC press LLC (2001)

Harlow, H.F., Mears, C.: The Human Model: Primate Perspectives. V. H. Winston and Sons, Washington (1979)

Harris, C.L., Ayçiçegi, A., Gleason, J.B.: Taboo words and reprimands elicit greater autonomic reactivity in a first language than in a second language. Applied Psycholinguistics 24, 561–579 (2003)

Hauser, M.D., Chomsky, N., Fitch, W.T.: The faculty of language: what is it, who has it, and how did it evolve? Science 298, 1569–1579 (2002)

Hilbert, D.: The foundations of mathematics. In: van Heijenoort, J. (ed.) From Frege to Gödel, p. 475. Harvard University Press, Cambridge (1928/1967)

von Humboldt, W.: Uber die Verschiedenheit des menschlichen Sprachbaues und ihren Einfluss auf die geistige Entwickelung des Menschengeschlechts. F. Dummler, Berlin (1836/1967); Also in Lehmann, W. P. (ed.) A Reader in Nineteenth Century Historical Indo-European Linguistics. Indiana University Press, Bloomington

Hurford, J.: The evolution of human communication and language. In: D'Ettorre, P., Hughes, D. (eds.) Sociobiology of communication: an interdisciplinary perspective, pp. 249–264. Oxford University Press, New York (2008)

Huron, D.: Ernest Bloch Lectures. University of California Press, Berkeley (1999)

Ilin, R., Perlovsky, L.I.: Cognitively Inspired Neural Network for Recognition of Situations. International Journal of Natural Computing Research 1(1), 36–55 (2010)

Jaynes, J.: The origin of consciousness in the breakdown of the bicameral mind. Houghton Mifflin Co., Boston (1976)

Joyce, D., Richards, L., Cangelosi, A., Coventry, K.R.: On the foundations of perceptual symbol systems: Specifying embodied representations via connectionism. In: Detje, F., Dörner, D., Schaub, H. (eds.) The Logic of Cognitive Systems. Proceedings of the Fifth International Conference on Cognitive Modeling, pp. 147–152. Universitätsverlag, Bamberg (2003)

Jung, C.G.: Psychological Types (1921); In the Collected Works, v.6, Bollingen Series XX. Princeton University Press, Princeton (1971)

Juslin, P.N., Sloboda, J.A.: Music and emotion: Theory and research. Oxford University Press, Oxford (2001)

Juslin, P.N., Västfjäll, D.: Emotional responses to music: The Need to consider underlying mechanisms. Behavioral and Brain Sciences 31, 559–575 (2008)

Justus, T., Hustler, J.J.: Fundamental issues in the evolutionary psychology of music: Assessing innateness and domain specificity. Music Perception 23, 1–27 (2003)

Kalman, R.E.: A new approach to linear filtering and prediction problems. Journal of Basic Engineering 82(1), 35–45 (1960)

Kant, I.: Critique of Judgment. In: Bernard, J.H. (tr.) Macmillan & Co., London (1790/1914)

Kant (1798/1974)

Kay, P.: An informal sketch of a formal architecture for construction grammar. Grammars 5, 1–19 (2002)

Kecman, V.: Learning and Soft Computing: Support Vector Machines, Neural Networks, and Fuzzy Logic Models (Complex Adaptive Systems). The MIT Press, Cambridge (2001)

Kosslyn, S.M.: Image and mind. Harvard University Press, Cambridge (1980)

Kosslyn, S.M.: Image and Brain. MIT Press, Cambridge (1994)

Kosslyn, S.M., Ganis, G., Thompson, W.L.: Neural foundations of imagery. Nature Reviews Neuroscience 2, 635–643 (2001)

Kovalerchuk, B., Perlovsky, L.I.: Dynamic Logic of Phenomena and Cognition. In: IJCNN 2009, Atlanta (2009); Submitted for Journal Publication

Kozma, R., Puljic, M., Perlovsky, L.: Modeling goal-oriented decision making through cognitive phase transitions. New Mathematics and Natural Computation 5(1), 143–157 (2009)

Kuhn, T.S.: The Structure of Scientific Revolutions. University of Chicago Press, Chicago (1962)

Lakoff, G.: Women, Fire, and Dangerous Things. What Categories Reveal about the Mind. University of Chicago Press, Chicago (1987)

Lakoff, G., Johnson, M.: Metaphors We Live By. University of Chicago Press, Chicago (1980)

Lakoff, G., Johnson, M.: Philosophy in the flesh: The embodied mind and its challenge to western thought. HarperCollins Publishers, New York (1999)

Langacker, R.W.: Foundations of Cognitive Grammar. In: Theoretical Prerequisites, vol. 1. Stanford University Press, Stanford (1987)

Langacker, R.W.: Language, Image and Symbol. Mouton de Gruyter, The Hague (1991)

Lakatos, I., Musgrave, A. (eds.): Criticism and the Growth of Knowledge: Proceedings of the International Colloquium in the Philosophy of Science, London, vol. 4 (1965)

Lerer, S.: Inventing English. Columbia University Press, Chichester (2007)

Levine, D.S., Perlovsky, L.I.: Neuroscientific insights on Biblical myths: Simplifying heuristics versus careful thinking: Scientific analysis of millennial spiritual issues. Zygon, Journal of Science and Religion 43(4), 797–821 (2008)

Levine, D.S., Perlovsky, L.I.: Emotion in the pursuit of understanding. International Journal of Synthetic Emotions 1(2), 1–11 (2010)

Levitin, D.J.: This is your brain on music: The science of a human obsession. Dutton, London (2006)

Levitin, D.J.: The world in six songs. Dutton, London (2008)

Linnehan, R., Mutz, C., Perlovsky, L.I., Weijers, B., Schindler, J., Brockett, R.: Detection of Patterns Below Clutter in Images. In: Int. Conf. on Integration of Knowledge Intensive Multi-Agent Systems, Cambridge, MA (2003)

Linnehan, R., Brady, D., Schindler, J., Perlovsky, L., Rangaswamy, M.: On the Design of SAR Apertures Using the Cram´er-Rao Bound. IEEE Transactions on Aerospace and Electronic Systems 43(1), 344–355 (2007)

Lorenz, K.: The foundations of ethology. Springer, New York (1981)

Luther, M.: Preface to Symphoniae jucundae. See W&T, p. 102 (1538)

Masataka, N.: The origins of language and the evolution of music: A comparative perspective (2008); Physics of Life Reviews, 6, 11–22 (2009); Lieberman, P.: Human language and our reptilian brain. Harvard University Press, Cambridge (2000)

Maimonides, M.: The Guide for the Perplexed (1190); Friedlander, M.(tr.) Unabridged Dover Edition (1956)

Marchal, B.: Theoretical Computer Science & the Natural Sciences. Physics of Life Reviews 2(3), 1–38 (2005)

Mayorga, R., Perlovsky, L.I. (eds.): Sapient Systems. Springer, London (2008)

Mayer, J.D., Salovey, P., Caruso, D.R.: Emotional Intelligence. New Ability or Eclectic Traits? American Psychologist 63(6), 503–517 (2008)

McCulloch, W.S.: Embodiments of Mind. The MIT Press, Cambridge (1988/1965)

McDermott, J., Houser, M.: The origins of music: Innateness, uniqueness, and evolution. Music Perception 23, 29–59 (2003)

McDermott, J.: The evolution of music. Nature 453, 287–288 (2008)

Meyer, R.K., Palmer, C., Mazo, M.: Affective and coherence responses to Russian laments. Music Perception 16(1), 135–150 (1998)

Meystel, A.M., Albus, J.S.: Intelligent systems: Architecture, design, and control. Wiley, New York (2001)

Minsky, M.L.: A Framework for Representing Knowlege. In: Whinston, P.H. (ed.) The Psychology of Computer Vision. McGraw-Hill Book, New York (1975)

Mitchell, T.M.: Machine Learning. WCB/McGraw-Hill, New York (1997)

Mithen, S.: A creative explosion? Theory of mind, language, and the disembodied mind of the Upper Paleolithic. In: Mithen, S. (ed.) Creativity in Human Evolution and Prehistory, pp. 165–191. Routledge, London and New York (1998)

Mithen, S.: The Singing Neanderthals. Harvard Univ. Press, Cambridge (2007)

Nevatia, R., Binford, T.O.: Description and recognition of curved objects. Artificial Intelligence 8(1), 77–98 (1977)

Nietzsche, F.: Untimely Meditations. In: Hollingdale (tr.) Cambridge Univ. Press, Cambridge (1876/1983)

Panksepp, J., Bernatzky, G.: Emotional sounds and the brain: The neuro-affective foundations of musical appreciation. Behavioural Processes 60, 133–155 (2002)

Patel, A.D.: Music, language, and the brain. Oxford Univ. Press, New York (2008); Parrott, W.: Emotions in Social Psychology. Psychology Press, Philadelphia (2001)

Peirce, C.S.: Logic as Semiotic: The Theory of Signs. In: Buchler (ed.) The Philosophical Writing of Peirce, Dover, NY (1955); Collected Papers V. II, Elements of Logic. In: Hartshorne, Weiss (eds.) Belknap, Cambridge, MA (1897, 1903)

Penrose, R.: Shadows of the Mind. Oxford University Press, Oxford (1994)

Perlovsky, L.I.: Scattering of Seismic Waves by Shear Modulus Fluctuations Due to Rigid-Nonrigid Phase Transition. In: Teorica, B.d.G. (ed.) Applicata, vol. XXVIII (109), pp. 3–13 (1986)

Perlovsky, L.I.: Multiple sensor fusion and neural networks. DARPA Neural Network Study (1987)

Perlovsky, L.I.: Cramer-Rao Bounds for the Estimation of Means in a Clustering Problem. Pattern Recognition Letters 8, 1–3 (1988a)

Perlovsky, L.I.: Frequency Estimates for Simple Oscillating Systems Under Random Forcing. In: Datta, B.N., et al. (eds.) Linear Algebra in Signals, Systems and Control. SIAM, Philadelphia (1988b)

Perlovsky, L.I.: Cramer-Rao Bounds for the Estimation of Normal Mixtures. Pattern Recognition Letters 10, 141–148 (1989)

Perlovsky, L.I.: Acoustic Critical Opalescence of the Transitional Under-Ice Layer in the Arctic. In: Teorica, B.d.G. (ed.) Applicata, vol. XXXIII (132), pp. 237–256 (1991)

Perlovsky, L.I.: Computational Concepts in Classification: Neural Networks, Statistical Pattern Recognition, and Model Based Vision. Journal of Mathematical Imaging and Vision 4(1), 81–110 (1994a)

Perlovsky, L.I.: A Model Based Neural Network for Transient Signal Processing. Neural Networks 7(3), 565–572 (1994b)

Perlovsky, L.I.: Cramer-Rao Bound for Tracking in Clutter and Tracking Multiple Objects. Pattern Recognition Letters 18(3), 283–288 (1997a)

Perlovsky, L.I.: Physical Concepts of Intellect. Proceedings of Russian Academy of Sciences 354(3), 320–323 (1997b)

Perlovsky, L.I.: Probabilistic Multi-Hypothesis Tracking. In: Streit, R.L. (ed.) Cramer-Rao Bound for Tracking in Clutter, pp. 77–84. NUWC Press, Newport (1998a)

Perlovsky, L.I.: Conundrum of Combinatorial Complexity. IEEE Trans. PAMI 20(6), 666–670 (1998b)

Perlovsky, L.I.: Beauty and mathematical Intellect. Zvezda 2000(9), 190–201 (2000) (Russian)

Perlovsky, L.I.: Neural Networks and Intellect: using model-based concepts. Oxford University Press, New York (2001) (3rd printing)

Perlovsky, L.I.: Physical Theory of Information Processing in the Mind: concepts and emotions. SEED On Line Journal 2(2), 36–54 (2002)

Perlovsky, L.I.: Statistical Limitations on Molecular Evolution. Journal of Biomolecular Structure & Dynamics 19(6), 1031–1043 (2002b)

Perlovsky, L.I.: Aesthetics and mathematical theories of intellect. Iskusstvoznanie 2(02), 558–594 (2002c) (Russian)

Perlovsky, L.I.: Integrating Language and Cognition. IEEE Connections, Feature Article 2(2), 8–12 (2004)

Perlovsky, L.I.: Autonomous Intelligent Systems: Agents and Data Mining. In: Gorodetsky, V., Liu, J., Skormin, V.A. (eds.) Evolving Agents: Communication and Cognition, pp. 37–49. Springer-Verlag, GmbH (2005)

Perlovsky, L.I.: Toward Physics of the Mind: Concepts, Emotions, Consciousness, and Symbols. Phys. Life Rev. 3(1), 22–55 (2006a)

Perlovsky, L.I.: Fuzzy Dynamic Logic. New Math. and Natural Computation 2(1), 43–55 (2006b)

Perlovsky, L.I.: Music – The First Principle. Musical Theatre (2006c), `http://www.ceo.spb.ru/libretto/kon_lan/ogl.shtml`

Perlovsky, L.I.: Artificial Cognition Systems. In: Loula, A., Gudwin, R., Queiroz, J. (eds.) Modeling Field Theory of Higher Cognitive Functions, pp. 64–105. Idea Group, Hershey (2007a)

Perlovsky, L.I.: Semiotics and Intelligent Systems Development. In: Gudwin, R., Queiroz, J. (eds.) Symbols: Integrated Cognition and Language, pp. 121–151. Idea Group, Hershey (2007b)

Perlovsky, L.I.: Aspects of Automatic Text Analysis (Festschrift in Honor of Burghard Rieger). In: Köhler, R., Mehler, A. (eds.) Neural Networks, Fuzzy Models and Dynamic Logic, pp. 363–386. Springer, Germany (2007c)

Perlovsky, L.I.: Neurodynamics of Higher-Level Cognition and Consciousness. In: Perlovsky, L.I., Kozma, R. (eds.) Neural Dynamic Logic of Consciousness: the Knowledge Instinct. Springer, Heidelberg (2007d), ISBN 978-3-540-73266-2

Perlovsky, L.I.: Cognitive high level information fusion. Information Sciences 177, 2099–2118 (2007e)

Perlovsky, L.I.: Evolution of Languages, Consciousness, and Cultures. IEEE Computational Intelligence Magazine 2(3), 25–39 (2007f)

Perlovsky, L.I.: The Mind vs. Logic: Aristotle and Zadeh. Society for Mathematics of Uncertainty, Critical Review 1(1), 30–33 (2007g)

Perlovsky, L.I.: Music and Consciousness. Leonardo Journal of Arts, Sciences and Technology 41(4), 420–421 (2008a)

Perlovsky, L.I.: Sapience, Consciousness, and the Knowledge Instinct (Prolegomena to a Physical Theory). In: Mayorga, R., Perlovsky, L.I. (eds.) Sapient Systems. Springer, London (2008b)

Perlovsky, L.I.: Language and Cognition. Neural Networks 22(3), 247–257 (2009a), doi:10.1016/j.neunet.2009.03.007

Perlovsky, L.I.: Language and Emotions: Emotional Sapir-Whorf Hypothesis. Neural Networks 22(5-6), 518–526 (2009b), doi:10.1016/j.neunet.2009.06.034

Perlovsky, L.I.: 'Vague-to-Crisp' Neural Mechanism of Perception. IEEE Trans. Neural Networks 20(8), 1363–1367 (2009c)

Perlovsky, L.I.: Musical emotions: Functions, origin, evolution. Physics of Life Reviews 7(1), 2–27 (2010a), doi:10.1016/j.plrev.2009.11.001

Perlovsky, L.I.: Intersections of Mathematical, Cognitive, and Aesthetic Theories of Mind. Psychology of Aesthetics, Creativity, and the Arts 4(1), 11–17 (2010b), doi:10.1037/a0018147

Perlovsky, L.I.: Neural Mechanisms of the Mind, Aristotle, Zadeh, & fMRI. IEEE Trans. Neural Networks 21(5), 718–733 (2010c)

Perlovsky, L.I.: The Mind is not a Kludge. Skeptic 15(3), 50–55 (2010d)

Perlovsky, L.I.: Science and Religion: Scientific Understanding of Emotions of Religiously Sublime (2010e) arXiv:1010.3059

Perlovsky, L.I.: Jihadism and Grammars. Comment to "Lost in Translation". Wall Street Journal (2010f), `http://online.wsj.com/community/leonid-perlovsky/activity`

Perlovsky, L.I.: Joint Acquisition of Language and Cognition. WebmedCentral BRAIN 1(10), WMC00994 (2010g), `http://www.webmedcentral.com/article_view/994`

Perlovsky, L.I.: Beauty and Art. Cognitive Function, Evolution, and Mathematical Models of the Mind. WebmedCentral PSYCHOLOGY 1(12), WMC001322 (2010h)

Perlovsky, L.I.: Physics of the Mind: Concepts, Emotions, Language, Cognition, Consciousness, Beauty, Music, and Symbolic Culture. WebmedCentral PSYCHOLOGY 1(12), WMC001374 (2010i)

Perlovsky, L.I.: Consciousness and Free Will, a Scientific Possibility Due To Advances in Cognitive Science. WebmedCentral PSYCHOLOGY 2011, 2(2), WMC001539 (2010j)

Perlovsky, L.I., Bonniot-Cabanac, M.-C., Cabanac, M.: Curiosity and pleasure. WebmedCentral PSYCHOLOGY 1(12), WMC001275 (2010)

Perlovsky, L.I., Chernick, J.A., Schoendorf, W.H.: Multi-Sensor ATR and Identification Friend or Foe Using MLANS. Neural Networks 8(7/8), 1185–1200 (1995)

Perlovsky, L.I., Coons, R.P., Streit, R.L., Luginbuhl, T.E., Greineder, S.: Application of MLANS to Signal Classification. Journal of Underwater Acoustics 44(2), 783–809 (1994)

Perlovsky, L.I., Deming, R.W.: Neural Networks for Improved Tracking. IEEE Trans. Neural Networks 18(6), 1854–1857 (2007)

Perlovsky, L.I., Deming, R.W.: Maximum Likelihood Joint Tracking and Association in a Strong Clutter without Combinatorial Complexity (2010), arXiv:1010.4236

Perlovsky, L.I., Goldwag, A.: The Grammatical Roots of Jihadism: How Cognitive Science Can Help Us Understand the War on Terror. WebmedCentral, Psychology 2(2), WMC001581 (2011)

Perlovsky, L.I., Ilin, R.: Neurally and Mathematically Motivated Architecture for Language and Thought. Special Issue "Brain and Language Architectures: Where We are Now?". The Open Neuroimaging Journal 4, 70–80 (2010a), `http://www.bentham.org/open/tonij/openaccess2.htm`

Perlovsky, L.I., Ilin, R.: Grounded Symbols in The Brain, Computational Foundations for Perceptual Symbol System. WebmedCentral PSYCHOLOGY 1(12), WMC001357 (2010b)

Perlovsky, L.I., Jaskolski, J.V.: Maximum Likelihood Adaptive Neural Controller. Neural Networks 7(4), 671–680 (1994)

Perlovsky, L.I., Kozma, R. (eds.): Neurodynamics of Higher-Level Cognition and Consciousness. Springer, Heidelberg (2007a), ISBN 978-3-540-73266-2

Perlovsky, L., Kozma, R.: Neurodynamics of Cognition and Consciousness. In: Perlovsky, L., Kozma, R. (eds.) Editorial - Neurodynamics of Cognition and Consciousness. Springer, Heidelberg (2007b), ISBN 978-3-540-73266-2

Perlovsky, L.I., Marzetta, T.L.: Estimating a Covariance Matrix from Incomplete Independent Realizations of a Random Vector. IEEE Trans. on SP 40(8), 2097–2100 (1992)

Perlovsky, L.I., Mayorga, R.: Sapient Systems. In: Mayorga, R., Perlovsky, L.I. (eds.) Preface. Springer, London (2008)

Perlovsky, L.I., McManus, M.M.: Maximum Likelihood Neural Networks for Sensor Fusion and Adaptive Classification. Neural Networks 4(1), 89–102 (1991)

Perlovsky, L.I., Plum, C.P., Franchi, P.R., Tichovolsky, E.J., Choi, D.S., Weijers, B.: Einsteinian Neural Network for Spectrum Estimation. Neural Networks 10(9), 1541–1546 (1997)

Perlovsky, L.I., Schoendorf, W.H., Burdick, B.J., Tye, D.M.: Model-Based Neural Network for Target Detection in SAR Images. IEEE Trans. on Image Processing 6(1), 203–216 (1997)

Perlovsky, L.I., Schoendorf, W.H., Garvin, L.C., Chang: Development of Concurrent Classification and Tracking for Active Sonar. Journal of Underwater Acoustics 47(2), 375–388 (1997)

Perlovsky, L.I., Schoendorf, W.H., Tye, D.M., Chang, W.: Concurrent Classification and Tracking Using Maximum Likelihood Adaptive Neural System. Journal of Underwater Acoustics 45(2), 399–414 (1995)

Perlovsky, L.I., Webb, V.H., Bradley, S.R., Hansen, C.A.: Improved ROTHR Detection and Tracking Using MLANS. AGU Radio Science 33(4), 1034–1044 (1998)

Perlovsky, L.I., Webb, V.H., Bradley, S.A., Hansen, C.: Probabilistic Multi-Hypothesis Tracking. In: Streit, R.L. (ed.) Improved ROTHR Detection and Tracking using MLANS, pp. 245–254. NUWC Press, Newport (1998)

Pinker, S.: How the mind works. Norton, New York (1997)

Pinker, S.: The Stuff of Thought: Language as a Window into Human Nature. Viking, NY (2007)

Plato: (IV BC). In: Cooper, L.(ed.) Oxford University Press, New York

Poincare. The beauty of a scientific theory, quotes (1908), `http://www.quotationspage.com/quote/26209.html`

Popper, K.: RC Series Bundle: Conjectures and Refutations: The Growth of Scientific Knowledge. Routeledge, New York (2002)

Purwins, H., Herrera, P., Grachten, M., Hazan, A., Marxer, R., Serra, X.: Computational models of music perception and cognition I: The perceptual and cognitive processing chain. Physics of Life Reviews 5, 151–168 (2008a)

Purwins, H., Herrera, P., Grachten, M., Hazan, A., Marxer, R., Serra, X.: Computational models of music perception and cognition II: Domain-specific music processing. Physics of Life Reviews 5, 169–182 (2008b)

Roberson, D., Davidoff, J., Braisbyb, N.: Similarity and categorization: neuropsychological evidence for a dissociation in explicit categorization tasks. Cognition 71, 1–42 (1999)

Russell, J.A.: Core Affect and the Psychological Construction of Emotion. Psychological Review 110(1), 145–172 (2003)

Russell, J.A., Barrett, L.F.: Journal of Personality and Social Psychology 76(5), 805-819 (1999)

Sapir, E.: Culture, Language and Personality: Selected Essays by Edward Sapir. University of California Press, Berkeley (1985)

Saussure, F.D.: Course In General Linguistics, p. 98. McGraw-Hill, New York (1916/1965)

Schacter, D.L., Addis, D.R.: The ghosts of past and future. Nature 445, 27 (2007)

Schulz, G.M., Varga, M., Jeffires, K., Ludlow, C.L., Braun, A.R.: Functional neuroanatomy of human vocalization: an H215O PET study. Cerebral Cortex 15(12), 1835–1847 (2005)

Seyfarth, R.M., Cheney, D.L.: Meaning and emotion in animal vocalizations. Ann. NY Academy Sci., 32–55 (December 2003)

Sloboda, J.A., Juslin, P.N.: Psychological perspectives on music and emotion. In: Juslin, P.N., Sloboda, J.A. (eds.) Music and Emotion: Theory and Research, pp. 71–104. Oxford University Press, Oxford (2001); Russell, B.: Introduction to Mathematical Philosophy, p.175. George Allen and Unwin, London (1919)

Simmons, W.K., Barsalou, L.W.: The similarity-in-topography principle: reconciling theories of conceptual deficits. Cogn. Neuropsychol. 20, 451–486 (2003)

Simmons, W.K., Stephan, B.H., Carla, L.H., Xiaoping, P.H., Barsalou, L.W.: fMRI evidence for word association and situated simulation in conceptual processing. Journal of Physiology - Paris 102, 106–119 (2008)

Singer, R.A., Sea, R.G., Housewright, R.B.: Derivation and Evaluation of Improved Tracking Filters for Use in Dense Multitarget Environments. IEEE Transactions on Information Theory IT-20, 423–432 (1974)

Spelke, E.: Principles of object perception. Cognitive Science 14, 29–56 (1990)

Spinoza, B.: Ethics. Penguin, New York (1677/2005)

Talmy, L.: How language structures space. Plenum Press, New York (1983)

Talmy, L.: The relation of grammar to cognition. In: Rudzka-Ostyn (ed.) Topics in Cognitive Linguistics, pp. 165–205. Benjamins, Amsterdam (1988)

Tomasello, M.: Constructing a Language, A Usage-Based Theory of Language Acquisition. Harvard University Press, Cambridge (2003)

Tikhanoff, V., Fontanari, J.F., Cangelosi, A., Perlovsky, L.I.: Language and cognition integration through modeling field theory: Category formation for symbol grounding. In: Kollias, S.D., Stafylopatis, A., Duch, W., Oja, E. (eds.) ICANN 2006. LNCS, vol. 4131, pp. 376–385. Springer, Heidelberg (2006)

Tupes, E.C., Cristal, R.E.: Personnel research lab Lackland AFB TX, Report. Recurrent personality factors based on trait ratings. Journal of Personality 60(2), 225–251 (1961)

Tversky, A., Kahneman, D.: Judgment under Uncertainty: Heuristics and Biases. Science 185, 1124–1131 (1974)

Trainor, L.: Innateness, learning, and the difficulty of determining whether music is an evolutionary adaptation. Music Perception 24, 105–110 (2004)

Trainor, L.: The neural roots of music. Nature 453(29), 598–599 (2008)

Trehub, S.E.: The developmental origins of musicality. Nature Neuroscience 6(7), 669–673 (2003)

Ungerer, F., Schmid, H.-J.: An introduction to cognitive linguistics. Pearson, New York (2006)

Vapnik, V.N.: Statistical Learning Theory. Wiley-Interscience, New York (1998)

Weiss, P., Taruskin, R.: Music in the Western World. Schirmer, Macmillan, New York (1984)

Wiener, N.: Extrapolation, Interpolation, and Smoothing of Stationary Time Series. Wiley, New York (1949)

Whorf, B.L.: Language, Thought, and Reality. MIT Press, Cambridge (1956)

Wikipedia. Linguistic relativity (2009a), http://en.wikipedia.org/wiki/Sapir-Whorf_Hypothesis

Wikipedia. Sapir-Whorf hypothesis (2009b), http://en.wikipedia.org/wiki/SWH

Winawer, J., Witthoft, N., Frank, M., Wu, L., Wade, A., Boroditsky, L.: Russian blues reveal effects of language on color discrimination. PNAS 104(19), 7780–7785 (2007), doi:10.1073/pnas.0701644104

Winston, P.H.: Artificial Intelligence. Addison-Wesley, Reading (1984)

Wu, L., Barsalou, L.W.: Perceptual simulation in conceptual combination: Evidence from property generation. Acta Psychologica (2009 (in print)

Yeh, W., Barsalou, L.W.: The situated nature of concepts. Am. J. Psychol. 119, 349–384 (2006)

Zadeh, L.A.: Fuzzy Sets. Information and Control 8, 338–352 (1965)